Applications of the Reflexive Game Theory: Advanced Topics

Applications of the Reflexive Game Theory: Advanced Topics

Sergey Tarasenko

With Foreword by Vladimir A. Lefebvre

Sergey Tarasenko
2016

Tarasenko, Sergey
Applications of the reflexive game theory: advanced topics / Sergey Tarasenko
Includes bibliographical references

Copyright © 2016 by Sergey Tarasenko

All rights reserved. This book or any portion thereof may not be reproduced or used in any manner whatsoever without the express written permission of the publisher except for the use of brief quotations in a book review or scholarly journal.

First Printing: 2016

ISBN 978-1-365-05268-2

Dedication

To my Family

Contents

Acknowledgements ix

Foreword x

Preface xiii

Introduction 1

Part I: Fundamentals of the Reflexive Game Theory 3

Chapter 1: Fundamentals of the Reflexive Game Theory 5
    1.1   Introduction: motivation to use RGT 5
    1.2   Formalism of the Reflexive Game Theory 6
    1.3   The Forward Task 9
    1.4   The Inverse Task 11
    1.5   Basic Control Schema of an Abstract Subject 23

Part II: Advanced Topics 27

Chapter 2: Introducing Robots in the Reflexive Game Theory 29
    2.1   Formalization of the Robotic Subjects 29
    2.2   Examples of Interactions between Humans and Robots: Robot Baby-Sitters. 32
    2.3   Examples of Interactions between Humans and Robots: A Rescue Robot and Mountain-Climbers 37
    2.4   Discussion and Conclusion 41

Chapter 3: Modeling Multi-stage Decision Making with the Reflexive Game Theory 43
    3.1   Introduction 43
    3.2   Model of two-stage decision making: formation of points of view 44
    3.3   A Model of a multi-stage decision making: set-up parameters of the final session 49

3.4 Modeling multi-stage decision making processes with the RGT 50
3.5 Discussion and conclusion 52

Chapter 4: Emotionally Colorful Reflexive Games 53
4.1 Emotions in the Reflexive Theory 53
4.2 Theory of Bipolar Constructs 54
4.3 Enriching Emotional Palette 56
4.4 Bridging the PAD and the RGT 57
4.5 Merging the RGT and the PAD 59
4.6 Emotionally Colorful Reflexive Games 60
4.7 Status of a Situation, Multi-stage Decisions and Emotional Reflexive Control 63
4.8 Concluding remarks 73

Chapter 5: Socializing Autonomous Units 77
5.1 Introduction 77
5.2 Resonate-and-Fire neurons: a Brief Overview 77
5.3 Building Communication System for Group to Make Groups of Autonomous Units 83
5.4 Discussion 92

Chapter 6: Modeling Social Dynamics 95
6.1 Social dynamics 95
6.2 Formation of relationships 95
6.3 Representation of the group structure 96
6.4 Transition between different group structures ... 99
6.5 Modeling Social Dynamics with Markov Stochastic Process 107

Notes 115

References 117

Acknowledgements

I would like to thank Professor Vladimir Lefebvre for a nearly five-year long fruitful collaboration, his valuable comments and comprehensive discussions.

Foreword

Sergey Tarasenko wrote a great book. He has not only added new facets to the Reflexive Game Theory, but also significantly expanded its application. Let me emphasize just a single point. The author is the first who conducted reflexive analysis of groups consisting of both humans and robots. The necessity to create such groups may appear already in a few decades, for example, in exploring the space or the ocean depths, so the studies on human-robot communications will become in demand. To effectively work and communicate with humans, robots must be equipped with a special cognitive model which would allow them to imitate mutual relations between robots and humans, robots and robots and even humans and humans. The author offers to use the reflexive game theory as the base for such model and develops interesting reflexive models of robots and humans, which will be helpful in creating mixed robot-human groups.

Vladimir A. Lefebvre
27 May 2016

Preface

This book is a result of five-year long work in the field of Reflexive Psychology together with Professor Vladimir Lefebvre. The material presented in this book contains some research results taken together in the form of a coherent framework. This framework is an essential extension of the basic Reflexive Game Theory proposed by Professor Vladimir Lefebvre.

Some concepts presented in this book such as Multi-stage Decision Making and Socialization of Autonomous Units can be used as standalone tools independent from the Reflexive Game Theory.

Other concepts such as Mixed Groups of Human and Robots, Emotionally Colorful Reflexive Games and Social dynamics are essential extension of the Reflexive Game Theory and should be heavily tested in the subsequent studies.

In this book, the author only introduces application recipes, while validity of these recipes and the Reflexive Game Theory itself is a task sensitive question, which should be individually answered by a particular researcher.

Introduction

This book introduces advanced topics of the Reflexive Game Theory (RGT). The book consists of two parts. Part I introduces fundamentals of the RGT, including formalism of the RGT, Forward Task and Inverse Task of the RGT. Part II illustrates advanced topics of the RGT. A word "advanced" implies here that the RGT is fused with numerous auxiliary methods to achieve particular practical goals.

The first topic is an extension of the RGT for the case of mixed groups of humans and robots. This requires formalization of robotic subjects in the RGT.

The second topic is Multi-stage Decision Making. The concept of a multi-stage decision making is thoroughly discussed together with underlying assumptions.

The third topic is Emotional Reflexive Games born as a fusion of the RGT with Pleasure-Arousal-Dominance (PAD) model. Fusion of the RGT and the PAD is not a simple straightforward task. To clearly explain the concepts behind and points to connect these concepts, essential background from psychological studies is included.

The fourth topic is focused on the issue of establishing an exemplar communication system for autonomous units (robots) to enable them to create groups and exchange their influences on each other.

The final topic is dedicated to the matter of social dynamics. The group dynamics is analyzed by means of Reflexive Markov process, where transition probabilities are generate by the RGT inference.

These topics are taken together in the form of a coherent framework, which is an essential extension of the basic Reflexive Game Theory.

Applications of the Reflexive Game Theory: Advanced Topics

Part I: Fundamentals of the Reflexive Game Theory

Chapter 1: Fundamentals of the Reflexive Game Theory

1.1 Introduction: motivation to use RGT

The social modeling are usually characterized with redundant information that makes it difficult to extract the crucial factors and make decision: "... many executives continue to feel overwhelmed (60% say they have more information than they "can effectively use") " (Hopkins, 2011).

Another problem in decision making is a lack of information.

Both redundancy and lack of information can cause a situation of uncertainty.

The perfect trade-off between lack and redundancy of information is when we have exactly as much information as we need. The questions "How much information do we need?" and "What information do we need?", can be actually answered. BUT in order to answer these questions, we have to clearly understand the entire structure of the problem, we are dealing with.

It appears to be that in many cases only some small amount of information is needed to model social decision making and evolution. Therefore, we need a simple minimal model to start with. In other words, we need some simple model, which is capable of extracting *the basic semantics* of the social situation and which is based on the minimal set of rules.

Most recently such a model has been introduced by Lefebvre (2009, 2010) in terms of Reflexive Game Theory (RGT). This theory allows to predict the decisions of subjects in the group. To do so, only the information about the relationships in the group and mutual influences is required.

The peculiarity of this theory is its scale invariance: the subjects (group members) in the group can be single individuals, groups of people, military units, cities, countries, etc.

In general, "vast amount of uncertainty makes human socio-cultural behavior studies (HSCBS) the realm of intuition and art as well as science" (Johnson, 2010).

The RGT, being a part of social science, follows this pattern. Although the basic theory is easy to understand, its application requires essential skills. Therefore, the majority of examples is a fusion of different techniques. We will illustrate numerous examples of application in Part 2.

1.2 Formalism of the Reflexive Game Theory

In this section we introduce basics of the RGT. The exhaustive description of the RGT and tutorial of RGT application have been presented by Lefebvre (2009, 2010). Here, we present a brief overview of the RGT enough to understand its basic concept and formulate the tasks solved in this books.

The RGT is designed to analyze decision of subjects within a certain group, when there exists a set of choice alternatives. Through interaction within a group, each subject should select an alternative of action. The complete set of all possible alternatives is represented in a form of Boolean algebra

For the purpose of analysis, each subject is assigned a unique *subject variable*. It is assumed that behavior of each subject in a group is determined by *reflexive function* Φ, which is a functions of mutual influences of subjects in a group (Lefebvre, 2009, 2010):

$$\Phi = \Phi(a_1,...,a_i,..., a_n), \qquad (1.1)$$

where n is a total number of subjects in a group; a_j, for any $j=1,...,n$, is an influence of subject a_j on a_i. Influences are also introduced in the form of possible alternatives.

Then selected alternatives of subject a_i are defined as solutions (if any) of *decision equation* (Lefebvre, 2009, 2010):

$$a_i = \Phi(a_1,...,a_i,..., a_n) \qquad (1.2)$$

In a group, each pair of subjects can be either in a conflict or in an alliance relationship. It is assumed that all subjects make influence on each other. Therefore, any group is represented in a form of a fully connected graph. Such graph is called a *relationship graph*. Each vertex of the graph corresponds to a single subject and is assigned a subject variable. The relationships are illustrated with graph ribs: the solid-line ribs correspond to an alliance, while dashed ones are considered as a conflict.

Each graph can be decomposable or non-decomposable. The decomposable relationship graphs (Lefebvre, 2009, 2010; Batchelder and Lefebvre, 1982) can be presented in the analytical form of a corresponding *polynomial*. Any relationship graph of three subjects is decomposable. To represent relationship graph in analytic form, alliance is considered to be conjunction (multiplication) operation (·), and conflict is defined as disjunction (summation) operation (+).

Here we analyze some abstract group of subjects. Consider three subjects a, b and c. Let subject a is in alliance with other subjects, while subjects b and c are in conflict (Fig. 1.1 a)).

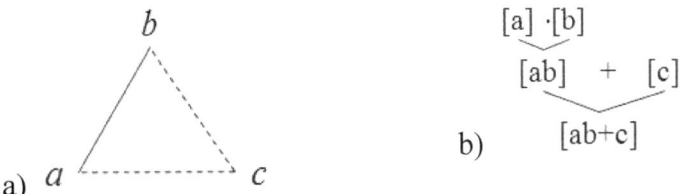

Fig. 1.1 Analysis of an abstract group: a) relationship graph, b) stratification tree.

The polynomial corresponding to this graph is $ab+c$.

Any polynomial can be stratified into sub-polynomials (Lefebvre, 2009, 2010) regarding either disjunction or conjunction operations. If stratification regarding conjunction is done first, then stratification regarding disjunction is done on the next step. Stratification procedure finalizes, when all sub-polynomials contain a single variable.

The result of stratification is the *Polynomial Stratification Tree* (PST). It has been proved that each non-elementary polynomial can be stratified in a unique way, i.e., each non-elementary polynomial has only one corresponding PST (Batchelder and Lefebvre, 1982).

PST for *ab+c* polynomial is presented in Fig. 1b).

The next step in a group analysis is to omit the branches of a PST and for each non-elementary polynomial write in the top right corner its sub-polynomials. The resulting tree-like structure is called a *diagonal form* (Lefebvre, 2009, 2010). Consider a diagonal form for the PST in Fig. 1.1 b):

$$\begin{array}{c} [a][b] \\ [ab] \quad + \quad [c] \\ [ab+c] \end{array}$$

Next we fold a diagonal form using exponential formula, defined by Lefebvre (2009, 2010):

$$P^W = P + \overline{W}, \qquad (1.3)$$

where $\overline{W}$ stands for the negation of W.

The folding using exponential formula is done as follows:

$$\begin{array}{c} [a][b] \\ [ab] \quad + \quad [c] \\ [ab+c] \qquad = ab+c. \end{array}$$

Then, we define reflexive function of each subject to be a result of a diagonal form folding operation:

$$\Phi(a,b,c) = ab+c \qquad (1.4)$$

Assuming, that all subjects have the same understanding of a group relationships, a reflexive function is the same for all subjects in a group. We define *decision equation* as follows:

$$x = ab+c, \qquad (1.5)$$

where x is any subject in a group.

Finally, we introduce a *canonical form* (Lefebvre, 2009, 2010) of a decision equation:

$$x = Ax+B\overline{x}, \qquad (1.6)$$

where A and B are some sets.

If $A \supseteq B$, then solution of a decision equations in a canonical form is given as interval $A \supseteq x \supseteq B$. Otherwise, there is no solution. If there is no solution, then a certain subject is considered to be in a frustration state.

Regarding the decision equation in the canonical form, is it possible to formulate two tasks: a *Forward task* and an *Inverse task*. We discuss these tasks in details in the following sections.

1.3 The Forward Task

Let us transform a general decision equation (1.5) into decision equations in a canonical form for each subject:

$$\begin{aligned} a &= (b+c)a + c\overline{a} \\ b &= (a+c)b + c\overline{b} \\ c &= c + ab\overline{c} \end{aligned} \qquad (1.7)$$

Variable in the left-hand side of each decision equation in system (1.7) is the variable of the equation. Other variables in the same equation are considered as influences on the subject from the other subjects.

Applications of the Reflexive Game Theory: Advanced Topics

The *Forward task* is formulated as a task to find the possible choices of a subject of interest, when the influences on him from other subjects are given.

After transformation of arbitral decision equation into its canonical form, the sets *A* and *B* are functions of other subjects' influences. In our example, the sets *A* and *B* are the functions of subject variables *b* and *c*: $A = A(b,c)$ and $B = B(b,c)$.

All the influences are presented in influence matrix (Table 1.1). The main diagonal of influence matrix contains the subject variables. The rows of the matrix represent influences of a given subject on other subjects, while columns represent the influences of other subjects on a given one.

Table 1.1 Influence Matrix

	a	b	c
a	a	{α}	{β}
b	{β}	b	{β}
c	{β}	{β}	c

The influence values are used in decision equations. Here we provide actual computations:

For subject *a*:
$$a = (\{\beta\}+\{\beta\})a + \{\beta\}\bar{a} \Rightarrow a = \{\beta\}a + \{\beta\}\bar{a} \Rightarrow a = \{\beta\}.$$

For subject *b*:
$$b = (\{\alpha\}+\{\beta\})b + \{\beta\}\bar{b} \Rightarrow b = b + \{\beta\}\bar{b}.$$

For subject *c*:
$$c = c + \{\beta\}\{\beta\}\bar{c} \Rightarrow c = c + \{\beta\}\bar{c}.$$

Equation for subject *a* turns into equality. This happens because $A(b,c) = B(b,c) = \{\beta\}$.

Equation for subject b turns into equality $b = b + \{\beta\}\overline{b}$. Therefore, $A = A(b,c) = 1 = \{\alpha,\beta\} \supseteq B = B(a,b) = \{\beta\}$. Thus there exists at least one solution from the interval $1 \supseteq b \supseteq \{\beta\}$.

Equation for subject c the same as for subject b. The solution belongs to the interval $1 \supseteq c \supseteq \{\beta\}$.

Therefore subjects b and c can choose any alternative from Boolean algebra, which contains alternative $\{\beta\}$. These alternatives are $1 = \{\alpha,\beta\}$ and $\{\beta\}$.

This concludes description of the Forward task.

1.4 The Inverse Task

In contrast to the Forward task, the *Inverse task* (Tarasenko, 2010) is formulated as a task to find all the simultaneous (or joint) influences of all the subjects together on the subject of interest that result in a choice of a particular alternative or some set alternatives. We call the subject of interest to be a *controlled subject*.

Let subject a be a controlled subject and a^* is a fixed value, representing an alternative or set of alternatives, which subjects b, c, etc. want subject a to choose. We call value a^* to be a *target choice*. By substituting subject variable a with fixed value a^*, we obtain the *influence equation*. If we substitute the subject variable a with fixed value a^* in the canonical form of the decision equation (eq. (1.3)), we obtain *the canonical form of the influence equation*:

$$a^* = A(b,c,...)a^* + B(b,c,...)\overline{a^*} \quad (1.8)$$

In contrast to the decision equation, which is equation of a single variable, the influence equation is the equation of multiple variables. The number of variables in influence equation can be less than $(n-1)$, where n is the total number of subjects in the group.

Therefore the variables that present in the influence equation are called *effective variables*[1].

The Inverse task is by definition[2] formalized as to find all the joint solutions of all subjects in the group, except for the controlled one, when the target choice is represented by interval $\chi_1 \supseteq a^* \supseteq \chi_2$, where χ_1 and χ_2 are some sets and $\chi_1 \supset \chi_2$. To solve the Inverse task, one should solve the system of influence equations:

$$\begin{cases} A(b,c,...) = \chi_1 \\ B(b,c,...) = \chi_2 \end{cases} \qquad (1.9)$$

If the target choice is a single alternative, then $\chi_1 = \chi_2 = a^*$.

The solutions of the system (1.9) are considered as reflexive control strategies. The solution of the Inverse task in particular is characterized from two points.

The first point is whether it is required to find the influence of a particular single subject or joint influences of a group of subjects. The second one is whether the target choice is represented as a single alternative or as an interval of alternatives. In the case, when target alternatives do not belong to a certain interval, a system of equations should be considered for each single alternative.

To illustrate these points, we continue analysis by using a group of subjects introduced in the previous sections: subjects a and b are in alliance with each other and in conflict with subject c.

Influence of a Single Subject vs. Joint Influences of a Group. First we consider example, when the influence of a single subject is required. Let subject b makes influence $\{\alpha\}$ and $a^* = \{\alpha\}$. Then we

---

[1] Consider polynomial $b(a+d)+c$. Decision equations for subject a and d are $a=b+c$ and $d=b+c$. The decision equations for subjects b and c are $b = b + c\overline{b}$ and $c = c + b\overline{c}$.

[2] We need a system of influence equations because solutions of the influence equation $a^* = A(b,c,...)a^* + B(b,c,...)\overline{a^*}$ itself only guaratee that the original decision equation $a = A(b,c,...)a + B(b,c,...)\overline{a}$ turns into true equality, but it is not guaranteed that these solutions are the only ones that turn decision equation into true equality.

need to find influences of a single subject c, which result in solution $a^* = \{\alpha\}$ of decision equation $a = ab+c$.

The canonical form of this influence equation for subject a is

$$a^* = (b+c)a^* + c\overline{a^*} \qquad (1.10)$$

Since $a^* = \{\alpha\}$, $\chi_1 = \chi_2 = \{\alpha\}$ and $b = \{\alpha\}$, we obtain a system of equations:

$$\begin{cases} \{\alpha\} + c = \{\alpha\} \\ c = \{\alpha\} \end{cases} \qquad (1.11)$$

Therefore, the straight forward solution of this system is $c = \{\alpha\}$.

This simple example illustrates the very gist of the *Inverse task* - to find the appropriate influences, which result in target choice.

Next, we consider that influence of subject b is not known. Therefore, we obtain system (1.12)

$$\begin{cases} b + c = \{\alpha\} \\ c = \{\alpha\} \end{cases} \qquad (1.12)$$

In this case, we need to find the values of variable b, which together with values of variable c, result in solution $a^* = \{\alpha\}$. In other words, we need to find all the pairs (b,c), resulting in solution $a^* = \{\alpha\}$.

Therefore, we run all the possible values of variable b and check if the first equation of the system (1.12) turns into true equality:

$b = 1$: $1 + \{\alpha\} = 1 \rightarrow 1 \neq \{\alpha\}$;
$b = \{\alpha\}$: $\{\alpha\} + \{\alpha\} = \{\alpha\} \rightarrow \{\alpha\} = \{\alpha\}$;
$b = \{\beta\}$: $\{\beta\} + \{\alpha\} = 1 \rightarrow 1 \neq \{\alpha\}$;
$b = 0$: $0 + \{\alpha\} = \{\alpha\} \rightarrow \{\alpha\} = \{\alpha\}$.

Therefore, out of four possible values, only two values $\{\alpha\}$ and 0 are appropriate. Thus, we obtain two pairs (b,c): $(\{\alpha\},\{\alpha\})$ and $(\{\alpha\},0)$.

A Single Target Alternative vs. Interval of Alternatives. In the previous examples we considered a target choice to be only a single alternative. Here we illustrate the case, when a target choice is an interval. Let $b = \{\beta\}$, and $1 \supseteq a^* \supseteq \{\alpha\}$. To find corresponding influences of subject c, we solve the system of equations:

$$\begin{cases} \{\beta\} + c = 1 \\ c = \{\alpha\} \end{cases} \quad (1.13)$$

Again, we instantly obtain the solution of this system: $c = \{\alpha\}$.

In this section, we have formulated the Inverse task in general and considered its particular formalization depending on the number of influences and what is the target choice. However, we do not have a method to solve any influence equation. Therefore, we solve this problem in the next section.

How to Solve an any Influence Equation. As an introduction, we consider the fundamental proposition, which will be the corner stone to solve the influence equations.

Proposition 1. Let P and Q be some abstract sets. Then
$P\overline{Q} + \overline{P}Q = 0 \Leftrightarrow P = Q$

Proof

Necessity. Let $P\overline{Q} + \overline{P}Q = 0$, then

$P\overline{Q} + \overline{P}Q = 0 \Rightarrow P\overline{Q} + \overline{P}Q + P = P \Rightarrow P + \overline{P}Q = P \Rightarrow$
$P(Q + \overline{Q}) + \overline{P}Q = P \Rightarrow Q + P\overline{Q} = P \Rightarrow (1 + \overline{P})Q + P\overline{Q} = P \Rightarrow$
$Q + \overline{P}Q + P\overline{Q} = P \Rightarrow Q = P$

Therefore if $P\overline{Q} + \overline{P}Q = 0$, then $P = Q$.

Sufficiency. Let $P = Q$, then $P\overline{Q} + \overline{P}Q = P\overline{P} + \overline{Q}Q = 0$.

Now let us consider the new type of equation:

$$A_1 x + B_1 \bar{x} = 0 \qquad (1.14)$$

Eq. (1.14) has solution if and only if

$$\overline{A_1} \supseteq x \supseteq B_1 \qquad (1.15)$$

Solving the Influence Equations. There are three operations defined on the Boolean algebra. They are conjunction ($\cdot$ or multiplication), disjunction (+ or summation) and negation $\bar{x}$, where x is a subject variable. The negation operation is unary operation, while other two operations are binary. Using combination of these three operations, we can compose any influence equation.

Since, it is obvious how to solve the equation including only unary operation, we discuss how to solve influence equations including a single binary operation.

For this purpose, we consider two abstract subject variables x_1 and x_2 and abstract alternative χ.

Since one bound of the solution intervals for eqs. (L1.1) and (L2.1) are functions of the second variable, we need to run all the possible values of the second variable in order to obtain all possible solutions of these equations in the form of pairs (x_1, x_2).

Next we consider several examples, illustrating application of Lemmas 1 and 2.

Applications of the Reflexive Game Theory: Advanced Topics

Lemma 1. The solution of equation
$$x_1 + x_2 = \chi, \quad (L1.1)$$
regarding variable x_i, where $i = 1, 2$, is given by the interval
$$\chi \supseteq x_i \supseteq (\overline{\chi} x_j + \overline{x_j}\chi), \quad (L1.2)$$
where $j = 1, 2; j \neq i$.

Proof

Consider eq. (L1.1). According to Proposition 1,
$P = x_1 + x_2$, $Q = \chi$,
$\overline{P} = \overline{x_1 + x_2} = \overline{x_1}\, \overline{x_2}$ and $\overline{Q} = \overline{\chi}$.

Therefore,
$$P\overline{Q} + \overline{P}Q = (x_1 + x_2)\overline{\chi} + \overline{x_1}\,\overline{x_2}\,\chi = x_1\overline{\chi} + x_2\overline{\chi} + \overline{x_1}\,\overline{x_2}\,\chi.$$

Consequently, we obtain eq. (L1.3):
$$x_1\overline{\chi} + x_2\overline{\chi} + \overline{x_1}\,\overline{x_2}\,\chi = 0 \quad (L1.3)$$

We solve eq. (L1.3) regarding variable x_1. First, we transform eq. (L1.3) into canonical form:
$$\overline{\chi}\, x_1 + (\overline{\chi}\, x_2 + \chi \overline{x_2})\overline{x_1} = 0 \quad (L1.4)$$

Therefore, the solution of eq.(L1.4) is given by the interval (L1.5):
$$\chi \supseteq x_1 \supseteq (\overline{\chi}\, x_2 + \chi \overline{x_2}) \quad (L1.5)$$

Since variables x_1 and x_2 are interchangeable and it is possible to solve eq.(L1.3) regarding variable x_2 as well, the general form of solution of eq.(L1.1) is the interval
$$\chi \supseteq x_i \supseteq (\overline{\chi}\, x_j + \chi \overline{x_j})$$
where $i = 1, 2$ and $j = 1, 2; j \neq i$.

Lemma 2. The solution of equation
$$x_1 x_2 = \chi \quad (L2.1)$$
regarding variable x_i, where $i = 1,2$, is given by the interval
$$\chi x_j + \overline{\chi}\, \overline{x_j} \supseteq x_i \supseteq \chi, \quad (L2.2)$$
where $j = 1,2;\ j \neq i$.

Proof.
Consider eq. (L2.1). According to Proposition 1,
$P = x_1 x_2,\ Q = \chi$,
$\overline{P} = \overline{x_1 x_2} = \overline{x_1} + \overline{x_2}$, and $\overline{Q} = \overline{\chi}$.

Therefore,
$$P\overline{Q} + \overline{P}Q = x_1 x_2 \overline{\chi} + (\overline{x_1} + \overline{x_2})\chi = x_2 \overline{\chi} x_1 + \overline{x_1}\chi + \overline{x_2}\chi.$$

Thus, we obtain eq.(L2.3):
$$x_2 \overline{\chi} x_1 + \overline{x_1}\chi + \overline{x_2}\chi = 0 \quad (L2.3)$$

We solve eq.(L2.3) regarding variable x_1. First, we transform eq. (ref{conEq}) into canonical form:
$$(\chi x_2 + \overline{\chi}\overline{x_2}) x_1 + \chi \overline{x_1} = 0 \quad (L2.4)$$

Since $\overline{\overline{\chi x_2} + \overline{\chi}\overline{x_2}} = \chi x_2 + \overline{\chi}\,\overline{x_2}$, the solution of eq.(L2.4) is given by the interval
$$\chi x_2 + \overline{\chi}\,\overline{x_2} \supseteq x_1 \supseteq \chi \quad (L2.5)$$

Since variables x_1 and x_2 are interchangeable and it is possible to solve eq.(L2.3) regarding variable x_2 as well, the general form of solution of eq.(L2.1) is the interval
$$(\chi x_j + \overline{\chi}\,\overline{x_j}) \supseteq x_i \supseteq \chi$$
where $i = 1,2$ and $j = 1,2;\ j \neq i$.

Example 1. For illustration, we solve equation $a^* = ba^*+c$. Consider $a^* = \chi$, $x_1 = ba^*$ and $x_2 = c$, we obtain the solution interval for variable $x_2 = c$: $\chi \supseteq c \supseteq (\chi \overline{\chi b} + \overline{\chi} \chi b)$. After simplification, we get interval (1.16):

$$\chi \supseteq c \supseteq \chi \overline{b} \qquad (1.16)$$

Next we consider examples with particular alternatives. Let it be alternative $\{\alpha\}$: $\chi = \{\alpha\}$.

The solution interval is then $\{\alpha\} \supseteq c \supseteq \{\alpha\}\overline{b}$. Since the lower bound of this interval is a function of variable b, to find all solutions of equation $a^* = ba^*+c$, we calculate value of expression $\{\alpha\}\overline{b}$ for all possible values of variable b (Table 1.2).

Table 1.2 Solutions of the influence equation $a^* = ba^*+c$

Values of b	$\{\alpha\}$	$\{\beta\}$	1	0
Pairs (b,c)	$(\{\alpha\},\{\alpha\})$	$(\{\beta\},\{\alpha\})$	$(1,\{\alpha\})$	$(0,\{\alpha\})$
	$(\{\alpha\},0)$		$(1,0)$	

To reassure that solutions are correct, we check that decision equation $a = ba+c$ turns into true equality for each obtained pair (b,c):

$(\{\alpha\},\{\alpha\})$: $\{\alpha\}\{\alpha\} + \{\alpha\} = \{\alpha\} \Rightarrow \{\alpha\} = \{\alpha\}$ is true;
$(\{\alpha\}, 0)$: $\{\alpha\}\{\alpha\} + 0 = \{\alpha\} \Rightarrow \{\alpha\} = \{\alpha\}$ is true;
$(\{\beta\}, \{\alpha\})$: $\{\alpha\}\{\beta\} + \{\alpha\} = \{\alpha\} \Rightarrow \{\alpha\} = \{\alpha\}$ is true;
$(1, \{\alpha\})$: $\{\alpha\}1 + \{\alpha\} = \{\alpha\} \Rightarrow \{\alpha\} = \{\alpha\}$ is true;
$(1, 0)$: $\{\alpha\}1 + 0 = \{\alpha\} \Rightarrow \{\alpha\} = \{\alpha\}$ is true;
$(0, \{\alpha\})$: $\{\alpha\}0 + \{\alpha\} = \{\alpha\} \Rightarrow \{\alpha\} = \{\alpha\}$ is true.

So far, we have illustrated how to solve the influence equation. We as well showed that the pairs (b,c) obtained by solving equation

$a* = ba*+c$ in accordance with Proposition 1 and Lemmas 1 and 2 are indeed solutions of this equation.

Example 2. We consider polynomial $a(b+c)$. The decision equation in the canonical form for subject b is:

$$b = b + (c+\bar{a})\bar{b} \qquad (1.17)$$

Therefore, the influence equation for subject b is eq.(1.18).

$$(c+\bar{a})\bar{\chi} + \chi = \chi \qquad (1.18)$$

First, we transform the left-hand side of eq. (1.18):
$(c+\bar{a})\bar{\chi} + \chi = c\bar{\chi} + \bar{a}\bar{\chi} + \chi =$
$c\bar{\chi} + \bar{a}\bar{\chi} + (c+\bar{a}+1)\chi = c + \bar{a} + \chi.$

Therefore, eq.(1.18) can be rewritten as follows:

$$c + \bar{a} + \chi = \chi \qquad (1.19)$$

Considering, $x_1 = c$ and $x_2 = \bar{a}+\chi$, we instantly obtain the solution interval of eq.(1.19):

$$\chi \supseteq c \supseteq (\bar{\chi}(\bar{a}+\chi) + \chi(\overline{a+\chi}) \Rightarrow \chi \supseteq c \supseteq (\bar{\chi}\bar{a} + \chi\bar{\chi}a).$$

Finally,

$$\chi \supseteq c \supseteq (\bar{\chi}\bar{a} + \chi\bar{\chi}a) \qquad (1.20)$$

Example 3. Next, we consider influence equation

$$ab+\chi = \chi \qquad (1.21)$$

Considering, $x_1 = ab$ and $x_2 = \chi$, we instantly obtain the solution interval $\chi \supseteq ab \supseteq (\chi\overline{\chi} + \overline{\chi}\chi)$ or

$$\chi \supseteq ab \supseteq 0 \qquad (1.22)$$

Therefore, in order to find all solutions of eq.(1.21), we need to solve the equations

$$ab = x, \qquad (1.23)$$

where x is any sub-set of set χ $(x \supset \chi)$.

Eq. (1.23) can be solved according to Lemma 2.

Example 4. As a final example, we again consider influence equation

$$a^* = (b+c)a^* + \overline{a^*} \qquad (1.10)$$

and show how application of Lemma 1 essentially simplifies its solution. We consider that $a^* = \{\alpha\}$, therefore $\chi_1 = \chi_2 = \{\alpha\}$. We obtain a system of influence equations:

$$\begin{cases} b+c = \{\alpha\} \\ c = \{\alpha\} \end{cases} \qquad (1.11)$$

From this system we obtain a single equation (1.24)

$$b + \{\alpha\} = \{\alpha\} \qquad (1.24)$$

According to Lemma 1, we instantly obtain the solution interval of eq.(1.24):

$$\{\alpha\} \supseteq b \supseteq 0 \qquad (1.25)$$

Thus, eq. (1.24) has two solutions: $b = \{\alpha\}$ and $b = 0$. Therefore the solution of system (1.11) consists of two pairs $(\{\alpha\},\{\alpha\})$ and $(0, \{\alpha\})$.

Example 4. Analysis of Extreme Cases 1: Frustration. In this example we analyze the situation, when subject can appear in frustration state, from the point of view of the inverse task.

Let's again consider the polynomial $a(b+c)$.

The decision equation for subject a is

$$a = (b+c)a + \overline{a} \qquad (1.26)$$

The solution interval of this equation is

$$(b+c) \supseteq a \supseteq 1 \qquad (1.27)$$

Influence equation (1.10) correspond to decision equation (1.26). We need to check which alternatives subject a can be convinced to choose. To do this, we consider the system of equations for each alternative.

Alternative $\{\alpha\}$:

$$\begin{cases} b+c = \{\alpha\} \\ 1 = \{\alpha\} \end{cases} \qquad (1.28)$$

Alternative $\{\beta\}$:

$$\begin{cases} b+c = \{\beta\} \\ 1 = \{\beta\} \end{cases} \qquad (1.29)$$

Alternative 0={}:

$$\begin{cases} b+c=0 \\ 1=0 \end{cases} \quad (1.30)$$

In systems of equations (1.28)-(1.30), the second equation is incorrect equality. Therefore these systems have no solution.

Alternative 1={α,β}:

$$\begin{cases} b+c=1 \\ 1=1 \end{cases} \quad (1.31)$$

The second equation in system (1.31) is correct equality. Therefore this system has solution. Consequently, out of four possible alternatives, subject *a* can actually choose only alternative 1={α,β}.

To find solutions, resulting in selection of the alternative 1={α,β}, we need to solve only eq. (1.32), since the second equation of system (1.31) is the true equality.

$$b+c=1 \quad (1.32)$$

According to Lemma 1, we instantly obtain the solution interval for eq.(1.32):

$$1 \supseteq b \supseteq \overline{c} \quad (1.33)$$

We calculate the pairs (b,c) for all possible values of variable c (Table 1.3).

Table 1.3 Solutions of the influence equation $a^* = (b+c)a^* + \overline{a^*}$

Values of c	$\{\alpha\}$	$\{\beta\}$	1	0
Pairs (b,c)	$(\{\beta\},\{\alpha\})$ $(1,\{\alpha\})$	$(\{\alpha\},\{\beta\})$ $(1,\{\beta\})$	$(0,1)$ $(\{\alpha\},1)$ $(\{\beta\},1)$ $(1,1)$	$(1,0)$

Therefore, the influence analysis of the decision equation $a = (b+c)a + \overline{a}$ shows that the only alternative that subject a can choose is alternative $1=\{\alpha,\beta\}$. The influence analysis provides us with the set (exhaustive list) of joint influences (b,c) resulting in selection of alternative $1=\{\alpha,\beta\}$. Therefore, if the pair of influences does not match any pair from this list, the decision equation has no solution and this results in frustration state.

1.5 Basic Control Schema of an Abstract Subject

Here we summarize all the presented material in the form of *Basic Control Schema of an Abstract Subject (BCSAS) in the RGT*. BCSAS is the fundamental schema of an abstract subject. The BCSAS is presented in Fig. 1.2.

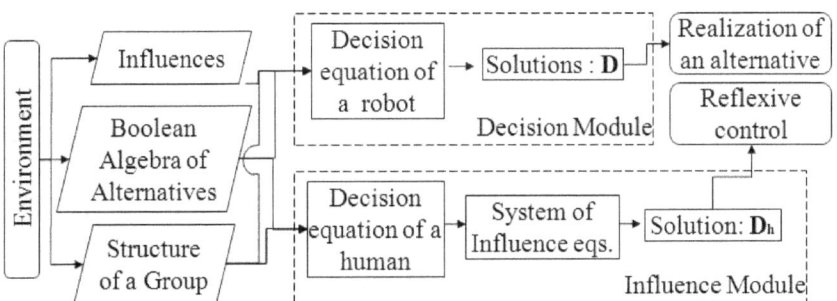

Fig. 1.2 The Basic Control Schema of an Abstract Subject (BSCAS).

The input comes from the environment and is formalized in the form of external Influences on the subject, the Boolean algebra of Alternatives and Structure of a Group.

Information about the Influences, Boolean algebra and Group Structure is propagated into the *Decision Module*.

The Decision Module implements solution of the Forward task. Therefore the output set D of the Decision Module is the set of possible alternatives, which subject can choose under the given influences.

The information about Boolean algebra and Group Structure is propagated into the *Influence Module*. The Influence Module solves the Inverse task. The output set D_h of the Influence Module is the set of the pairs $(\chi, Z_\chi)_x$, where χ is the target alternative, the set Z_χ is the set of all the joint influences, resulting in selection of the target choice; and x represents a subject variable. Each $(\chi, Z_\chi)_x$ represents a reflexive control strategy.

Therefore, the decision to put a subject into frustration state is justified if it is impossible to make subject x choose the target alternative χ, i.e., if for pair $(\chi, Z_\chi)_x$ set $Z_\chi = \{\}$, and subject x should not choose any other alternative except for the target one.

The alternatives χ with corresponding non-empty sets Z_χ are included into the set D_h. Here we introduce set Z_h to store the non-empty sets Z_χ.

The schema of the algorithm for extracting sets D_h and Z_h is presented in Fig. 1.3. First the sets D_h and Z_h are empty: $D_h = \{\}$ and $Z_h = \{\}$. The algorithm reads the set of pairs $(\chi, Z_\chi)_x$ and stores it in array Pairs(M), where M is a counting variable, N is the total number of pairs. Then it is checked for each pairs from array Pairs whether set Z_χ is empty: $Z_\chi == \{\}$? . If 'yes', the algorithm increments counting variable M ($M = M+1$) and proceeds to the next pair from array Pairs. If 'no', then alternative chi is included into the set D_h ($D_h = D_h + \chi$), set D_h is saved, the set Z_χ is included into set Z_h ($Z_h = Z_h + Z_\chi$) and set Z_h is saved. The process is run while $M \le N$.

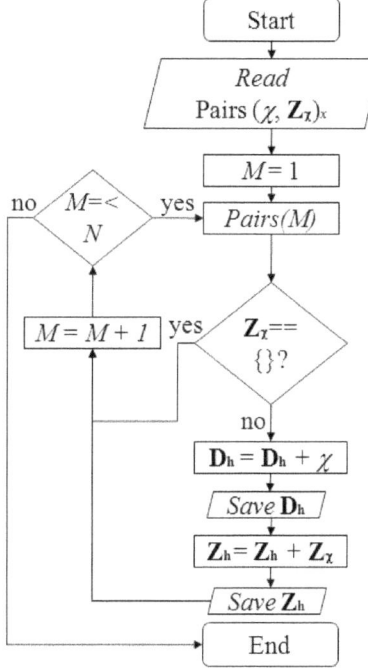

Fig. 1.3 The block schema for extracting sets D_h and Z_h.

In this iterative algorithm, we separately store the alternatives chi, which can be chosen by a certain subject, in the set D_h and the joint influences Z_χ, which result in selection of alternative χ, in the set Z_h.

Therefore, we should modify the schema of Influence Module in BCSAS as follows. We present elaborated schema, where sub-module "Solution: D_h" is accompanied with sub-module "Solution: Z_h". Together these sub-modules are included into the "Solution" sub-module.

This concludes an overview of the RGT and description of tasks within the scope of the general theory.

Part II: Advanced Topics

Chapter 2: Introducing Robots in the Reflexive Game Theory

2.1 Formalization of the Robotic Subjects

We start the Part II with application of the RGT to the mixed groups of humans and robots. This application has been first proposed by Tarasenko (2011). The goal of the robots in mixed groups of humans and robots is to refrain human subject from choosing actions, which might harm human's physical or mental health.

It is considered by default that robots are managed by a control system, which consists of at least three modules. The Module 1 implements RGT decision making or RGT inference. The Module 2 contains the rules, which refrain robot from making a harm to human beings and filters out alternatives containing possibly harmful or risky actions. The Module 3 predicts the choice of each human subject using the RGT and suggests possible reflexive control strategies.

The Modules 1 and 3 are inherited from the BCSAS. They correspond to Decision Module and Influence Module of the BCSAS (Fig. 1.2), respectively. Therefore all the properties and meaning of outputs of the Modules 1 and 3 are the same as the ones for Decision and Influence modules, respectively.

The Module 2 is a new module, which is intrinsic for robotic agents. This module is responsible for extraction of only harmless or non-risky alternatives for human subject.

We apply Asimov's Three Laws of robotics (Asimov, 1942), as a basis for Module 2.

The First Law: A robot may not injure a human being or, through inaction, allow a human being to come to harm.

The Second Law: A robot must obey any orders given to it by human beings, except where such orders would conflict with the First Law.

The Third Law: A robot must protect its own existence as long as such protection does not conflict with the First or Second Law.

The interaction of Modules 1 and 2 is performed in the Interaction Module 1. The interaction of Modules 3 and 2 is implements in the Interaction Module 2.

The Boolean algebra is filtered according to Asimov's laws in Module 2. The output of Module 2 is set U of approved alternatives. This data is then propagated into interaction modules.

The output of the Module 1 is set D of alternatives, which robot has to choose under the given joint influences. In the Interaction Module 1, the conjunction of sets D and U is performed: $D \cap U = DU$. If set DU is not empty set, this means that there are approved alternatives among the alternatives that robot should choose in accordance with the joint influences. Therefore, robot can implement any alternative from the set DU. If set DU is empty, this means that under given joint influences robot cannot choose any approved alternative, therefore robot will choose an alternative from set U. This is how the Interaction Module 1 works.

The output of the Module 3 contains sets D_h and Z_h. The goal of the robot is to refrain human subjects from choosing harmful alternatives. This can be done by convincing human subjects to choose alternatives from the set U. First, we check, whether D_h contains any approved alternative. We do so by performing conjunction of sets D_h and U: $D_h \cap U = D_h U$.

If set $D_h U$ is not empty, then it means that it is possible to make a human subject choose some alternative without harmful actions. Therefore, robot should choose the corresponding reflexive control strategy from the set Z_h. However, if set $D_h U$ is empty, a robot has to find the reflexive control strategy that will make human subject to select approved alternative from set U. For this purpose, we construct set Z_U, by including all the joint influences Z_χ for approved alternatives: Z_χ in $Z_U \Leftrightarrow \chi$ in U.

Next we check whether set Z_U is empty. If set Z_U is empty this means it is impossible to convince a human subject to choose alternative with harmless action. Therefore, the only option of reflexive control in this case is to put this subject into frustration state.

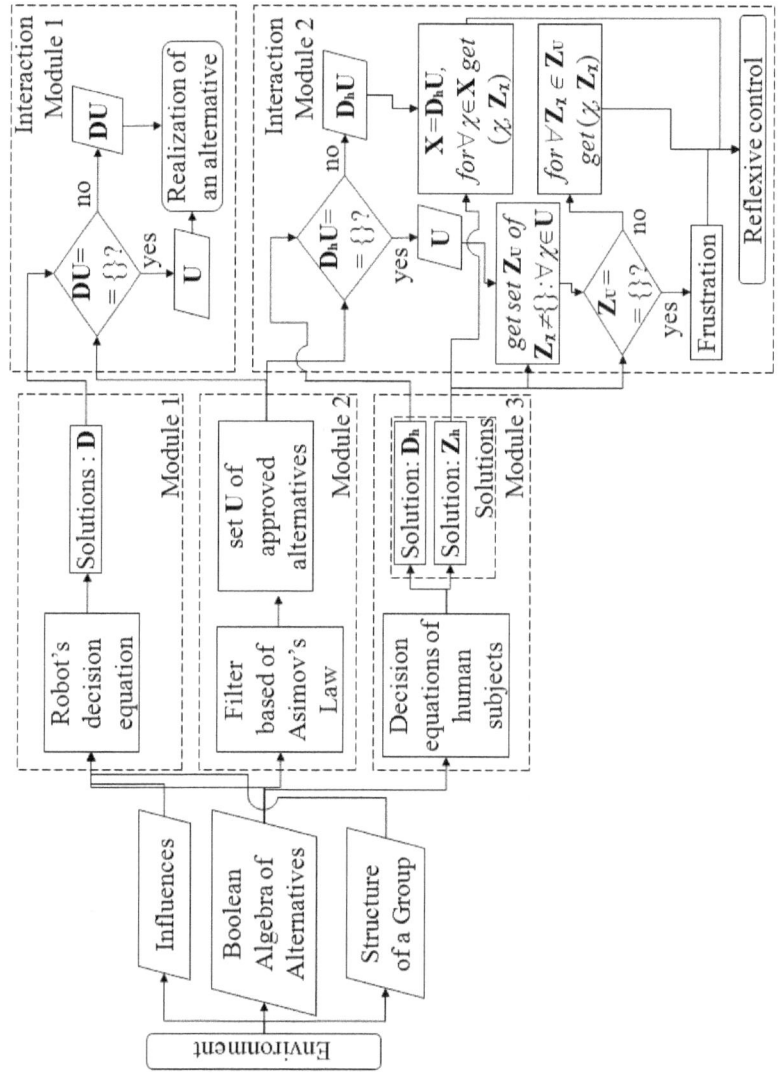

Fig. 2.1 The Basic Control Schema of a Robotic Agent (BCSRA).

However, if set Z_U is not empty, this means that there exist at least one reflexive control strategy that results in selection of alternative from the set of the approved (non-risky) ones.

The Modules 1, 2 and 3 together with Interaction Modules 1 and 2, represent Basic Control Schema for Robotic Agents (BCSRA). Therefore, the BCSRA inherits the entire structure of the BCSAS and augments it with Module 2 of Asimov's Laws together with Interaction Modules 1 and 2.

The original schema of robot's control system has been presented by Tarasenko (2011). The BCSRA is extended version of the original schema. The BCSRA provides comprehensive approach of how Forward and Inverse tasks are solved in the robot's "mind".

Thus, in this section we have presented the formalization of robotic agent in the RGT. We have outlined the specific features of robotic agents, which distinguish them from other subjects. Furthermore, we have provided detailed explanation of the how the Forward and Inverse tasks are solved in the framework of control system (BCSRA) of robots.

Next, we proceed with consideration of sample situations of interactions between humans and robots. We are going to elaborate two examples, presented in the previous study by Tarasenko (2011), of how robots in the mixed groups can make humans refrain from risky actions. We discuss the application of BCSRA and provide explicit derivation of reflexive control strategies, which have been applied in the study by Tarasenko (2011).

2.2 Examples of Interactions between Humans and Robots: Robot Baby-Sitters.

In the original set-up described by Tarasenko (2011), robots have to play a part of baby-sitters. A mixed group of human and robots consists of two kids and two robots. Each robot is looking after a particular kid.

Kids have finished some game and now are choosing between two actions: 1) "to compete climbing the high tree" (action α); and 2) "to play with a ball" (action β).

Therefore the Boolean algebra of alternatives includes four elements:

1) {α} is "to climb a tree";
2) {β} is "to play with a ball";
3) 1={α, β} means that a kid is hesitating what to do; and
4) 0 = {} means "to take a rest".

It is assumed that that each kid is alliance with his robot and in conflict with another kid and his robot.

Originally, kids were assigned variables subjects a and c, while robots were subjects b and d. The relationship graph is presented in Fig. 2.2.

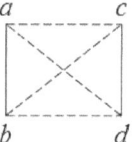

Fig. 2.2 The relationship graph for robots baby-sitters examples.

The remaining part of this example is focused on the solutions of the Inverse task regarding harmless alternatives {β} and {}.

We conduct the analysis regarding kid a. This analysis can be further extended for kid c in the similar manner.

The decision equation for kid a is

$$a = ab+cd \qquad (2.1)$$

First we transform it into canonical form:

$$a = (b+cd)a + cd\overline{a} \qquad (2.2)$$

Solve the Inverse Regarding a Single Alternative. Next we consider system of influence equations:

$$\begin{cases} b+cd = \chi & (2.3) \\ cd = \chi & (2.4) \end{cases}$$

where alternative χ in U.
Considering eq.(2.4), eq.(2.3) is transformed into equation (2.5):

$$b + \chi = \chi \qquad (2.5)$$

The solution of eq.(2.5) directly follows from Lemma 1:

$$\chi \supseteq b \supseteq 0 \qquad (2.6)$$

Therefore for $\chi = \{\beta\}$ and $\chi = \{\}$ the solutions are $\{\beta\} \supseteq b \supseteq 0$ and $b = 0$, respectively. Eq.(2.4) is solved according to Lemma 2:

$$\chi d + \overline{\chi}\,\overline{d} \supseteq c \supseteq \chi \qquad (2.7)$$

First Consider $\chi = \{\beta\}$. Then $\{\beta\}d + \{\alpha\}\overline{d} \supseteq c \supseteq \{\beta\}$. By varying values of variable d, we obtain all the pairs (c,d):

$d = 1$: $\{\beta\} \supseteq c \supseteq \{\beta\} \Rightarrow c = \{\beta\}$.
Therefore the solution is pair $(\{\beta\},1)$.

$d = 0$: $\{\alpha\} \supseteq c \supseteq \{\beta\}$.
Since $\{\alpha\} \cap \{\beta\} = \{\}$, there is no solution.

$d = \{\alpha\}$: $0 \supseteq c \supseteq \{\beta\}$.
Since $\{\beta\} \supset \{\}$, there is no solution.

$d = \{\beta\}$: $1 \supseteq c \supseteq \{\beta\}$.

Therefore there are two solutions $(1, \{\beta\})$ and $(\{\beta\},\{\beta\})$.
Therefore equation $cd = \{\beta\}$ has three solutions
$(\{\beta\},1)$, $(1, \{\beta\})$ and $(\{\beta\},\{\beta\})$.

Thus, we have solved both equations from system (2.3-2.4). The solutions of this system are the triplets (b,c,d) of joint influences, which are all possible combinations of solutions of both equations. Since there are two solution of eq.(2.3) and three solutions of eq.(2.4), there are six triplets (b,c,d) in total:
$(0, \{\beta\},1)$ and $(\{\beta\},\{\beta\},1)$;
$(0, 1, \{\beta\})$ and $(\{\beta\}, 1, \{\beta\})$;
$(0, \{\beta\}, \{\beta\})$ and $(\{\beta\},\{\beta\},\{\beta\})$.

Next we consider the case, when $\chi = 0 = \{\}$. Then interval (2.7) turns into interval $\overline{d} \supseteq c \supseteq 0$. We obtain pairs (c,d) for all values of variable d:

$d = 1$: $\overline{1} \supseteq c \supseteq 0 \Rightarrow c = 0$.
Thus, there is only one solution $(0,1)$.

$d = 0$: $1 \supseteq c \supseteq 0$.
Thus, there are four solutions $(1,0)$, $(\{\alpha\},0)$, $(\{\beta\},0)$ and $(1,0)$.

$d = \{\alpha\}$: $\{\beta\} \supseteq c \supseteq 0$.
Thus, there are four solutions $(\{\beta\},\{\alpha\})$ and $(0, \{\alpha\})$.

$d = \{\beta\}$: $\{\alpha\} \supseteq c \supseteq 0$.
Thus, there are four solutions $(\{\alpha\},\{\beta\})$ and $(0, \{\beta\})$.

In total, equation $cd = 0$ has 9 solutions. Therefore system (2.3-2.4) also has 9 solutions in the form of triplets (b,c,d):
$(0,1,0)$, $(0,0,0)$, $(0,0, \{\alpha\})$,
$(0,0, \{\beta\})$, $(0,0,1)$, $(0, \{\alpha\}, \{\beta\})$,

(0, {α},0), (0, {β},{α}) and (0, {β},0).

Solve the Inverse Regarding an Interval of Alternatives. We have considered two cases, when both upper and lower bounds of the interval of decision equation equal to the same alternative.

Now we discuss a new situation, when variable a should take not a single value, but several values. In this case, we should find the joint influences (b,c,d) that result in selection of either alternative {β} or {}. Since, {β}⊃{}, we need to find all the triplets (b,c,d), resulting in the solution of decision equation as interval {β}⊇a⊇{}.

Therefore, we need to solve the following system of equations:

$$\begin{cases} b + cd = \{\beta\} & (2.8) \\ cd = 0 & (2.9) \end{cases}$$

The eq.(2.8) turns into equality $b = \{\beta\}$, and we need to solve eq.(2.9). However, this equation has been already solved in the previous example. Therefore we obtain the solutions of the system (2.8-2.9):

({β},1,0), ({β},0,0), ({β},0, {α}),
({β},0,{β}), ({β},0,1), ({β},{α},{β}),
({β},{α}, 0), ({β},{β},{α}) and ({β},{β}, 0).

Comparing solutions of all three systems of influence equation, we can see that there are four solutions ({β},{β},{β}) and ({β},{},{β}); ({β},1, {β}) and ({β},{α},{β}). The first pair of solution results in choice of only alternative {β}, while second pair of solutions results in selection of either alternative {β} or alternative {}. These four solutions together illustrate that if $b = d = \{\beta\}$, it is guaranteed that regardless of influence of kid c, kid d will choose either of approved alternatives.

By analogy, we can see that among solutions of system (2.3-2.4) with $\chi = 0 = \{\}$, there are four solutions (0,1,0), (0,0,0),

(0,{α},0) and (0,{ β},0). Therefore, if $b = d = 0$, kid a will choose alternative 0={} regardless of influence of kid c.

These two examples of binding variables b and d were considered in the *Scenario 1* and *Scenario 2* of sample situation with robots baby-sitters, originally presented by Tarasenko (2011).

Summarizing the results of this section, we have shown that robots can successfully control kids' behavior by refraining them from doing risky actions. The basis of this control is entirely based on the proposed schema of robot's control system. We have analyzed all the possible reflexive control strategies by solving three systems of influence equation: two systems regarding a single alternative and one system regarding the interval of alternatives. Therefore, we have shown how the Inverse Task can be effectively solved by our proposed algorithm in situation similar to the real conditions.

2.3 Examples of Interactions between Humans and Robots: A Rescue Robot and Mountain-Climbers

This example is also adopted from Tarasenko (2011). In the original set-up, two mountain climbers and one rescue robot were considered. One of the climbers (subject b) needs help. The second climber (subject a) wants to rescue climber b (action α). The second option is to send a rescue robot (action β).

In this case, inaction (0={}) is inappropriate solution, thus the set U of approved alternatives for robot includes only alternative {β}. The goal of the robot is to refrain climber a from choosing alternative {α} and perform rescue mission by itself. All three subjects are assume to be in an alliance. The corresponding polynomial is [abc]. Folding a diagonal form results in alternative 1={α,β} for all the subject:

$$[abc]^{[a][b][c]} = 1$$

Therefore, it is impossible to influence any subject, and this group is uncontrollable. According to BCSRA, this means that set D_hU is empty, therefore robot needs to find a way to put a controlled subject into frustration state.

In the original example by Tarasenko (2011), robot c managed to change a relationships with climber b and changes the group structure to polynomial $a(b+c)$. Decision equation for subject a in this group is

$$a = (b+c)a + \bar{a} \qquad (2.10)$$

and corresponding decision interval is

$$(b+c) \supseteq a \supseteq 1 \qquad (2.11)$$

If $b+c \subset 1$, then there is no solution and climber a appears in a frustration situation. Here, climber a cannot make any decision and, thus, cannot do anything. This situation is equivalent to inaction. However, such situation is acceptable, because inaction alternative $(0=\{\})$ is considered to be harmful only for the rescue robot.

In order to put subject a into frustration state, the reflexive control strategy should NOT be selected from the list of solutions:
$(\{\beta\},\{\alpha\})$; $(1,\{\alpha\})$; $(\{\alpha\},\{\beta\})$; $(1,\{\beta\})$;
$(0,1)$; $(\{\alpha\},1)$; $(\{\beta\},1)$; $(1,1)$ and $(1,0)$.

Here we provide two examples of such joint influences (b,c):
$(\{\alpha\}, \{\alpha\}) \Rightarrow (\{\alpha\}+\{\alpha\}) = \{\alpha\} \supseteq 1$ and $(\{\beta\}, \{\}) \Rightarrow (\{\beta\}+\{\}) = \{\beta\} \supseteq 1$.

Here, we confirm that rescue robot can complete the mission itself. The decision equation for robot c is

$$c = c + (b+\bar{a})\bar{c} \qquad (2.12)$$

and corresponding decision interval is

$$1 \supseteq c \supseteq (b + \bar{a}) \qquad (2.13)$$

Here we analysis all 16 possible reflexive control strategies (a,b) that climbers can apply to robot c.

Examples with empty set DU.
For $(0,b)$, there will be the same situation regardless of value of variable b : $1 \supseteq c \supseteq (b + \bar{0}) \Rightarrow 1 \supseteq c \supseteq (b+1) \Rightarrow c = 1$.

For $(a,1)$, there will be the same situation regardless of value of variable a : $1 \supseteq c \supseteq (1+\bar{a}) \Rightarrow c = 1$.

For $(\{\alpha\},\{\alpha\})$:
$1 \supseteq c \supseteq (\{\alpha\} + \overline{\{\alpha\}}) \Rightarrow 1 \supseteq c \supseteq (\{\alpha\} + \{\beta\}) \Rightarrow c = 1$.

For $(\{\beta\},\{\beta\})$:
$1 \supseteq c \supseteq (\{\beta\} + \overline{\{\beta\}}) \Rightarrow 1 \supseteq c \supseteq (\{\beta\} + \{\alpha\}) \Rightarrow c = 1$.
Therefore in these cases set $D = \{\{\alpha,\beta\}\}$.

Next we consider other pairs (a,b).
$(1,\{\alpha\})$:
$1 \supseteq c \supseteq (\{\alpha\} + \bar{1}) \Rightarrow 1 \supseteq c \supseteq \{\alpha\}$.
Here set $D = \{\{\alpha,\beta\}, \{\alpha\}\}$.

$(\{\beta\},\{\alpha\})$:
$1 \supseteq c \supseteq (\{\alpha\} + \overline{\{\beta\}}) \Rightarrow 1 \supseteq c \supseteq \{\alpha\}$.
Here set $D = \{\{\alpha,\beta\}, \{\alpha\}\}$.

$(\{\beta\},0)$:
$1 \supseteq c \supseteq (0 + \overline{\{\beta\}}) \Rightarrow 1 \supseteq c \supseteq \{\alpha\}$.
Therefore, set $D = \{\{\alpha,\beta\}, \{\alpha\}\}$.

Since $U = \{\{\beta\}\}$, $DU = \{\}$ for all the cases considered above, therefore robot will choose alternative $\{\beta\}$ from the set U.

Examples with non-empty set DU
Consider the following pairs (a,b):

$(1,\{\beta\})$:
$1 \supseteq c \supseteq (\{\beta\} + \overline{1}) \Rightarrow 1 \supseteq c \supseteq \{\beta\}$.
Therefore, set $D = \{\{\alpha,\beta\}, \{\beta\}\}$.

$(1,0)$: $1 \supseteq c \supseteq (0 + \overline{1}) \Rightarrow 1 \supseteq c \supseteq 0$.
Thus, set $D = \{\{\alpha,\beta\}, \{\alpha\}, \{\beta\}, \{\}\}$.

$(\{\alpha\},\{\beta\})$: $1 \supseteq c \supseteq (\{\beta\} + \overline{\{\alpha\}}) \Rightarrow 1 \supseteq c \supseteq (\{\beta\}$.
Thus, set $D = \{\{\alpha,\beta\}, \{\beta\}\}$.

$(\{\alpha\},0)$: $1 \supseteq c \supseteq (0 + \overline{\{\alpha\}}) \Rightarrow 1 \supseteq c \supseteq \{\beta\}$.
Thus, set $D = \{\{\alpha,\beta\}, \{\beta\}\}$.

Since $U = \{\{\beta\}\}$, $DU = \{\{\beta\}\}$ for all the cases considered above, the robot will choose alternative $\{\beta\}$ from the set DU.

Thus, we have shown that under all 16 reflexive control strategies (a,b), robot c can choose the alternative $\{\beta\}$, which is "to perform a rescue mission by itself". Therefore robot will choose alternative $\{\beta\}$ regardless of the joint influences (a,b) of the climbers.

The discussed example illustrates how robot can transform uncontrollable group into controllable one by manipulating the relationships in the group. In the controllable group by its influence on the human subjects, robot can refrain climber a from risky action to rescue climber b. Robot achieves its goal by putting climber a into frustration state, in which climber a cannot make any decision. On the other hand, set U of approved alternatives guarantees that robot itself will choose the option with no risk for humans and implement it regardless of climber's influence.

This example illustrates application of two-stage decision making. On the first stage, robot changes a structure of a group. On the second-stage the final decision is made (Chapter 3).

Therefore, in this section we have illustrated robot's ability to refrain human being from risky actions and to perform these risky actions itself. This proves that our approach achieves both goals of robotic agent: 1) to refrain people from risky actions and 2) to perform risky actions by itself regardless of human's influences.

2.4 Discussion and Conclusion

Summarizing, we outline the most important results presented in Chapters 1 and 2. First of all, we have introduced the Inverse task and developed the ultimate methods to solve it. We have provided a comprehensive tutorial to the Reflexive Game Theory. The tutorial contains the detailed description of Forward and Inverse tasks together with method to solve them.

We have proposed control schemas for both abstract subject (BCSAS) and robotic agent (BCSRA). These schemas were especially designed to incorporate solution of Forward and Inverse Task, thus providing us with autonomous units (individuals, subjects, agents) capable of making decisions in the human-like manner. We have shown that robotic agents based on BCSRA can be easily included into the mixed groups of humans and robots and effectively serve their fundamental goals (refraining humans from risky actions and, if needed, perform the risky actions itself).

Chapter 3: Modeling Multi-stage Decision Making with the Reflexive Game Theory[3]

3.1 Introduction

The Reflexive Game Theory (RGT) allows to predict choices of subjects in the group. To do so, the information about a group structure and mutual influences between subjects is needed.

The group structure means the set of pair-wise relationships between subjects in the group. These relationships can be either of alliance or conflict type. The mutual influences are formulated in terms of elements of Boolean algebra, which is build upon the set of universal actions. The elements of Boolean algebra represent all possible choices. The mutual influences are presented in the form of Influence matrix.

In general, RGT inference can be presented as a sequence of the following steps:
1) formalize choices in terms of elements of Boolean algebra of alternatives;
2) presentation of a group in the form of a fully connected relationship graph, where solid-line and dashed-line ribs (edges) represent alliance and conflict relationships, respectively;
3) if relationship graph is decomposable, then it is represented in the form of polynomial: alliance and conflict are denoted by conjunction (·) and disjunction (+) operations;
4) diagonal form transformation (build diagonal form on the basis of the polynomial and fold this diagonal form);
5) deduct the decision equations;

---

[3] Material of this chapter has been first presented in Tarasenko (2013)

6) input influence values into the decision equations for each subject.

Let us call the process of decision making in a group to be a session. Therefore, RGT inference is a single session.

3.2 Model of two-stage decision making: formation of points of view

This chapter is dedicated to the matter of setting mutual influences in a group by means of *reflexive control* (Lefebvre, 1965).

The influences, which subjects make on each other, could be considered as a result of a decision making session previous to *ultimate decision making* (final session). The influences of this type we would call *set-up influences*. The set-up influences are intermediate result of the overall decision making process. The term set-up influences is related to the influences, which are used during the final session, only.

Consequently, the overall decision making process could be segregated into two stages. Let the result of such discussion (decision making) be a particular decision regarding the matter under consideration. We assume that actual decision making regarding the matter of interest (final session - Stage 2) is preceded by a *preliminary session* (Stage 1), which is about a decision making regarding the influences (points of view), which each subject will support during the final session. Such overall decision making process we call two-stage decision making process. The general schema of the two-stage decision making is presented in Fig.1.

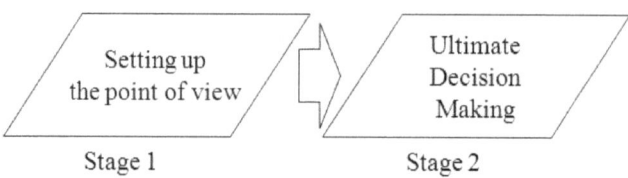

Fig. 3.1 The general schema of the two-stage decision making.

To illustrate such model we consider a simple example.

Example 1. Let director of some company has a meeting with his advisors. The goal of this meeting is to make decision about marketing policy for the next half a year. The background analysis and predictions of experts suggest three distinct strategies: aggressive (action α), moderate (action β) and soft (action γ) strategies.

The points of view of director and his advisors are formulated in terms of Boolean algebra of alternatives. A term *point of view* implies that a subject makes the same influences on the others. Director supports moderate strategy ($\{\beta\}$), the first and the second advisors are supporting aggressive strategy ($\{\alpha\}$), and the third advisor defends the idea of soft strategy ($\{\gamma\}$). The matrix of initial influences is presented in Table 3.1.

Table 3.1 Matrix of initial points of view (influences) used in Example 1.

	a	b	c	d
a	a	$\{\alpha\}$	$\{\alpha\}$	$\{\alpha\}$
b	$\{\alpha\}$	b	$\{\alpha\}$	$\{\alpha\}$
c	$\{\beta\}$	$\{\beta\}$	c	$\{\beta\}$
d	$\{\gamma\}$	$\{\gamma\}$	$\{\gamma\}$	d

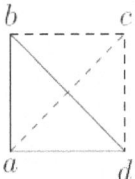

Fig. 3.2 Relationship graph for a director-advisors group.

Let director b is in a conflict with all his advisors, but his advisors are in alliance with each other. Variable c represents the Director, variables a, b and d correspond to the 1st, the 2nd and the 3rd advisor, respectively.

Applications of the Reflexive Game Theory: Advanced Topics

The relationship graph is presented in Fig. 3.2. Polynomial $abd+c$ corresponds to this graph.

After diagonal form transformation the polynomial does not change:

$$[abd+c] \quad \begin{matrix}[abd] \\ \end{matrix} \quad \begin{matrix}[a][b][d] \\ +[c]\end{matrix} = abd + c$$

Then we obtain four decision equation and their solutions (decision intervals) (Table 3.2):

Table 3.2 Decision equation and their solutions for Example 1.

Subject	Decision Equations	Decision Intervals
a	$a = (bd+c)a + c\bar{a}$	$(bd+c) \supseteq a \supseteq c$
b	$b = (ad+c)b + c\bar{b}$	$(ad+c) \supseteq b \supseteq c$
c	$c = c + abd\bar{c}$	$1 \supseteq c \supseteq abd$
d	$d = (ab+c)d + c\bar{d}$	$(ab+c) \supseteq d \supseteq c$

Next we calculate the decision intervals by using information from the influence matrix:

For subject a:
$(bd+c) \supseteq a \supseteq c \Rightarrow (\{\alpha\}\{\gamma\}+\{\beta\}) \supseteq a \supseteq \{\beta\} \Rightarrow a = \{\beta\}$;

For subject b:
$(ad+c) \supseteq b \supseteq c \Rightarrow (\{\alpha\}\{\gamma\}+\{\beta\}) \supseteq b \supseteq \{\beta\} \Rightarrow b = \{\beta\}$;

For subject c:
$1 \supseteq c \supseteq abd \Rightarrow 1 \supseteq c \supseteq \{\alpha\}\{\alpha\}\{\gamma\} \Rightarrow 1 \supseteq c \supseteq 0 \Rightarrow c = c$;

subject d:
$(ab+c) \supseteq d \supseteq c \Rightarrow (\{\alpha\}\{\alpha\}+\{\beta\}) \supseteq d \supseteq \{\beta\} \Rightarrow$
$\{\alpha,\beta\} \supseteq d \supseteq \{\beta\}$.

Therefore, after the preliminary sessions, the point of view of the subjects have changed. Director has obtained a freedom of choice, since he can choose any alternative: $1 \supseteq c \supseteq 0 \Rightarrow c = c$.

At the same time, the 1st and the 2nd advisors support moderate strategy: $a = b = \{\beta\}$. Finally, the 3rd advisor now can choose between points of view $\{\alpha,\beta\}$ (aggressive or moderate strategy) and $\{\beta\}$ (moderate strategy) : $\{\alpha,\beta\} \supseteq d \supseteq \{\beta\}$.

Thus the points of view of the 1st and the second advisors are definite, while the point of view of 3rd advisor is probabilistic.

Next we calculate choice of each subject during the final session considering the influences resulting from the preliminary session. The matrix of set-up influences is presented in Table 3.3. The intervals in matrix imply that a subject can choose either of alternatives from the given interval as a point of view.

Table 3.3 The matrix of set-up influences for Example 1.

	a	b	c	d
a	a	$\{\beta\}$	$\{\beta\}$	$\{\beta\}$
b	$\{\beta\}$	b	$\{\beta\}$	$\{\beta\}$
c	$1 \supseteq c \supseteq 0$	$1 \supseteq c \supseteq 0$	c	$1 \supseteq c \supseteq 0$
d	$1 \supseteq d \supseteq \{\beta\}$	$1 \supseteq d \supseteq \{\beta\}$	$1 \supseteq d \supseteq \{\beta\}$	d

Subject a:
$d = \{\alpha,\beta\}$:
$(bd+c) \supseteq a \supseteq c \Rightarrow \{\beta\}\{\alpha,\beta\}+c \supseteq a \supseteq c \Rightarrow \{\beta\}+c \supseteq a \supseteq c$.

$d = \{\beta\}$:
$(bd+c) \supseteq a \supseteq c \Rightarrow \{\beta\}\{\beta\}+c \supseteq a \supseteq c \Rightarrow \{\beta\}+c \supseteq a \supseteq c$.

Subject b:
$d = \{\alpha,\beta\}$:
$(ad+c) \supseteq b \supseteq c \Rightarrow (\{\beta\}\{\alpha,\beta\}+c) \supseteq b \supseteq c \Rightarrow \{\beta\}+c \supseteq b \supseteq c$.

$d = \{\beta\}$:
$(bd+c) \supseteq a \supseteq c \Rightarrow \{\beta\}\{\beta\}+c \supseteq a \supseteq c \Rightarrow \{\beta\}+c \supseteq b \supseteq c$.

Subject c:
$d = \{\alpha,\beta\}$:
$1 \supseteq c \supseteq abd \Rightarrow 1 \supseteq c \supseteq \{\beta\}\{\beta\}\{\alpha,\beta\} \Rightarrow 1 \supseteq c \supseteq \{\beta\}$.

$d = \{\beta\}$:
$1 \supseteq c \supseteq abd \Rightarrow 1 \supseteq c \supseteq \{\beta\}\{\beta\}\{\beta\} \Rightarrow 1 \supseteq c \supseteq \{\beta\}$.

Subject d:
$(\{\beta\}\{\beta\}+c) \supseteq d \supseteq c \Rightarrow (\{\alpha\}\{\alpha\}+\{\beta\}) \supseteq d \supseteq \{\beta\} \Rightarrow \{\beta\}+c \supseteq d \supseteq c$.

Now we compare the results of a single session with the ones of the two-stage decision making.

The single session case has been considered above. Therefore if the final decision have been made after the single session, then the 3rd advisor would be able to choose alternative $\{\alpha,\beta\}$ and realize action α. This option implies that each advisor is responsible for a particular part of the entire company and can take management decisions on his own.

Next we consider the decision made after the two-stage decision making. In such a case, regardless of influence of the 3rd advisor (subject d), advisors a and b influence is defined by the interval $\{\beta\}+c \supseteq x \supseteq c$, where x is either a or b variable.

Thus, if director is inactive ($c=0$), subjects a and b can choose either moderate strategy ($\{\beta\}$) or make no decision ($0=\{\}$). The same is true for subject d.

If the director makes influence $\{\beta\}$, then all the advisors will choose alternative $\{\beta\}$.

The director himself can choose from the interval $1 \supseteq c \supseteq \{\beta\}$ after the final session. This means that director c can choose any alternative containing action β. Thus, occasionally the director can realize his initial point of view as moderate strategy.

This example illustrates how using the two-stage decision making it is possible to make one's opponents choose the one's point of view. Meanwhile a person interested in such reflexive control can still sustain the initial point of view.

The obtained results are applicable in both cases when 1) only a director makes a decision; or 2) the decision are made individually by each subject.

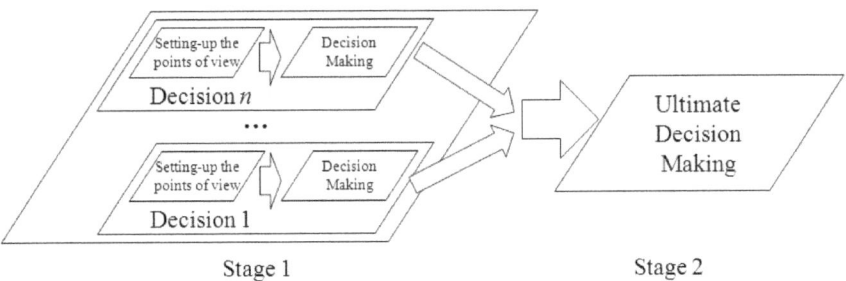

Fig. 3.3 Multi-stage decision making model.

3.3 A Model of a multi-stage decision making: set-up parameters of the final session

Now we consider the two-stage model in more details. In the considered example, during the preliminary session only the decision regarding the influences has been under consideration. In general case, however, before the final session has begun, there can be made decisions regarding any parameters of the final session. Such parameters include by are not limit to:
1) group structure (relationships between subjects in a group);
2) points of view;
3) decision to start a final session (a time when the final session should start), etc.

We call the decisions regarding a single parameter to be *consecutive decisions*, and decisions regarding distinct parameters to be *parallel*.

Therefore, during the first stage (before the final session) it is possible to make multiple decisions regarding various parameters of the final session. These decisions could be both parallel and consecutive ones. Such model of decision making we call multi-stage process of decision making (Fig.3.3).

3.4 Modeling multi-stage decision making processes with the RGT

Next we consider realization of multi-stage decision making with RGT.

Example 2: Change a group structure. Considering the subject from Example 1, we analyze the case when director wants to exclude the 3rd advisor from the group which would make the final decision.

In such a case there is a single action – 1 – "to exclude subject d from the group". Then Boolean algebra of alternatives includes only two elements: 1 и 0. Furthermore, it is enough that director just raise a question to exclude subject d from a group and make influence 1 on each subject: if $c = 1$, then $a=1$, $b=1$ and $d=1$ (Table 3.2). Thus the decision to exclude subject d from the group would be made automatically (Fig. 3.4).

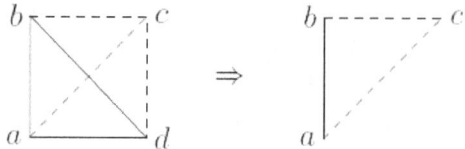

Fig. 3.4 Exclusion of a subject d from a group.

Example 3: Realization of a multi-stage decision making. Let the first decision discussed during the first stage is a decision regarding influences (points of view). The next decision was about exclusion of a subject *d* from the group. Thus, during the first step the formation of points of view has been implemented, then the structure of a group was changed. Therefore the group, which should make a final decision is described by polynomial $ab+c$. The decision equations and their solutions are presented in Table 3.4.

Table 3.4 Decision equations and decision interval for Example 3.

Subject	Decision Equation	Decision Interval
a	$a = (b+c)a + c\overline{a}$	$(b+c) \supseteq a \supseteq c$
b	$b = (a+c)b + c\overline{b}$	$(a+c) \supseteq b \supseteq c$
c	$c = c + ab\overline{c}$	$1 \supseteq c \supseteq ab$

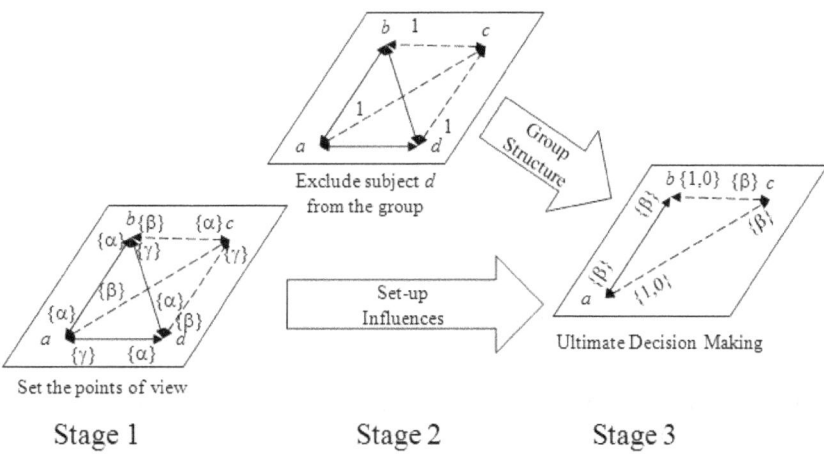

Fig. 3.5 Illustration of multi-stage decision making process. The influences are indicated by the arrow-ends of the ribs. The actual influence is presented near the arrow-end.

We consider that the point of view cannot change without preliminary session. Therefore we assume that the points of view do not change after the change of group structure. Therefore, during the final session the subjects would make the set-up influences derived from the preliminary session: subjects a and b will make influences $\{\beta\}$ and $1 \supseteq c \supseteq \{\beta\}$, respectively.

Such process is introduced in Fig. 3.5. During the 1st stage (the first step), the points of view of subjects have been formed. On the 2nd stage (the second step), the decision to exclude subject d from a group has been made. Finally, during the 3rd stage the final decision regarding the marketing strategy has been made.

3.5 Discussion and conclusion

This chapter introduces the two-stage and multi-stage decision making processes. During the first stage the decisions regarding the parameters of the final session are considered. The intermediate decisions are made during the preliminary sessions, while the final (ultimate) decision is made during the final session.

This chapter shows how before the final decision making the intermediate decision regarding parameters of the final session can be made and how the overall process of decision making could be represented as a sequence of decision making sessions. This approach enables complex decisions, which involve numerous parameters.

The important feature of the multi-stage decision making is that during the preliminary sessions subjects can convince other subjects to accept their own point of view. Therefore other subjects can be convinced to make decisions beneficial for a particular one. Such approach also allows to distribute the responsibility between all the members of the group, who make the final decision.

The result presented in this chapter allows to extend the scope of applications of RGT to model multi-stage decision making processes. Therefore it becomes possible to perform scenario analysis of various future trends and apply reflexive control to a project management.

Chapter 4: Emotionally Colorful Reflexive Games

4.1 Emotions in the Reflexive Theory

The *Reflexion Theory* (Lefebvre 1965, 1982, 2001) considers subjects, capable of making decisions. The concept of inner feelings (or emotions) is intrinsic for the Reflexion Theory. It is considered that emotions as a natural form of regulatory processes and reflection of deeper inner feelings in some way can be developed by collective subjects (groups) as well.

Previously, in the Reflexion Theory, the emotional experience of a subject has been modeled with a single dimension being *Pleasure*. The opposite of Pleasure is *Displeasure*. The feeling of displeasure can be directed towards the self, when we feel *self-disgraced* or *guilty*. If the displeasure is directed towards some external situation, for example, relationship with someone else, we suffer. Finally, if the source of one's displeasure is another person, the one's negative reaction can be characterized as *condemnation* or *blame*.

Grading the feeling of pleasure on the normalized scale [0,1], we consider 0 value to refer to the most displeased condition and 1 value to reflect the extreme pleasure. The 0 and 1 values are the negative and positive poles, respectively, on this scale. Since we have measured the *Pleasure (P)*, the *disPleasure (disP)* can be calculated as:.

$$disP = 1 - P \qquad (4.1)$$

In terms of the Reflexion Theory, *Pleasure* variable corresponds to the *ethical status* (Lefebvre 1982, 2001). Considering eq. (4.1), the displeasure can have a psychological meaning of *self-*

guiltiness, *suffering* or *condemnation*, depending on the source displeasure (Lefebvre 1982, 2001).

Imagine two human subjects *a* and *b*. Let the self in this case is subject *a*, another person is subject *b* and a *situation* is conflict between *a* and *b*. Then the self-guiltiness u, suffering v and condemnation ω are described by eqs. (4.21.4):

$$u = 1 - |a| \quad (4.2)$$
$$v = 1 - |a \ conflict \ b| \quad (4.3)$$
$$\omega = 1 - |b| \quad (4.4)$$

where notation $|\cdot|$ implies absolute value of the ethical status[4] estimated by subject *a*, i.e., $|a|$ is the self-estimated ethical status of subject *a*, $|a \ conflict \ b|$ is the ethical status of conflict situation between subject, and $|b|$ is the ethical status of subject *b*, which is estimated by subject *a* (Lefebvre 1982, 2001).

In the briefly discussed example, we have operated by only a single variable of the human internal world, i.e., pleasure. From our analysis it follows that this variable itself can be used to roughly describe the emotional experience of a person. But is this single variable enough?

4.2 Theory of Bipolar Constructs

This chapter is focused of extending RGT for the case, when subjects are enriched by ability to have emotional experience. This emotional aspect is simulated with 3d model, incorporating *Pleasure*, *Arousal* and *Dominance* components (*PAD* model) (Russel and Mehrabian, 1977; Mehrabian, 1996) of subject's behavior.

Next we consider the fundamental psychological theory first assumed the way of perception through the prism of bipolar deci-

---

[4] For detailed explanation of the concept of *ethical status* and its estimation please refer to Lefebvre (2001,Chapters 5, 6 and 7). The matter of self-guiltiness, suffering and condemnation is discussed in Lefebvre (2001, Chapters 6).

sions, and which has intrinsic connections with most of consequent psychological theories.

Kelly (1955) developed the experimental approach for assembling the list of personal constructs and a list of one's images of other people. The assumption behind is that each person has one's own unique system of dichotomous (bipolar) constructs, which serve as special axes for "projecting" self and other persons. Most of the constructs may be mapped onto the scale of "good-bad". The poles, the constructs are represented with, are of negative ("bad") and positive ("good") types. Hereafter, the theory proposed by Kelly will be referred as *Theory of Bipolar Constructs*. According to Kelly, these poles should be chosen with equal probability. However the subsequent psychological experiments proved this suggestion wrong.

In earlier experiments, it was shown by Benjafield and Adams-Webber (1976), Adams-Webber (1978), Benjafield and Green (1978), Shalit (1980), Osgood and Richards (1973) and Osgood (1979) that under great variety of experimental conditions the positive pole is chosen with the frequency 0.62.

The persistent appearance of the frequency 0.62 resulted in two ways. First, the reasonable question "What is the true value of the probability this frequency represents?". Second, it was suggested by Lefebvre (1985)(p. 291) that

"... human cognition has a special mechanism for modeling self and others. It works as a universal `inner computer' and creates the core of the images that later are `dressed' and `colored' with nuances. The constant 0.62 is a characteristic of this `computer'."

Lefebvre explained appearance of constant 0.62 by constructing a theoretical model of the bipolar choice (Lefebvre, 1985, 2006). This model predicted that the true value of the frequency 0.62 is probability 0.618..., which is inverted Golden Ratio ϕ (ϕ = 1.618...).

Therefore, the Golden Ratio, as being a characteristic of the *inner computer*, is related to the *Implicit Primordial Knowledge (IPK)* or *Pre-Knowledge* and is a key element of the *Basic Processing Algorithms* (Tarasenko et al., 2006).

Now we return to the Theory of Bipolar Constructs. It had strong impact on subsequently conducted psychological researches. The research about the emotions was not an exception.

4.3 Enriching Emotional Palette

One dimensional model of emotions has been proposed by Lefebvre to illustrate human inner feelings of self-guiltiness, suffering and condemnation (Lefebvre, 1982, 2001).

The semantic differential approach originally proposed by Osgood et al. (1957) considers three dimensions to characterize the person's personality (inner feelings). These dimensions are *Evaluation*, *Activity* and *Potency*.

This approach was further tested from the point of view of emotions. Russel and Mehrabian (1977) proved that the entire spectra of emotions can be described by the 3-dimensional space spanned by *Pleasure* (*Evaluation*), *Arousal* (*Activity*) and *Domination* (*Potency*) axes.

For every dimension, the lower and upper bounds (ends) are recognized as negative and positive poles, respectively. Consequently, the negative pole can be described by negative adjectives, and positive one - by positive adjectives.

It was shown by Russel and Mehrabian (1977) that these three dimensions are not only necessary dimensions for an adequate description of emotions, but they are also sufficient to define all the various emotional states. In other words, the *Emotional state* can be considered as a function of *Pleasure*, *Arousal* and *Dominance*.

Mehrabian (1996) proposed Pleasure-Arousal-Dominance (PAD) model of Emotional Scales. The emotional states defined as combinations of ends from various dimensions are presented in Fig. 4.1.

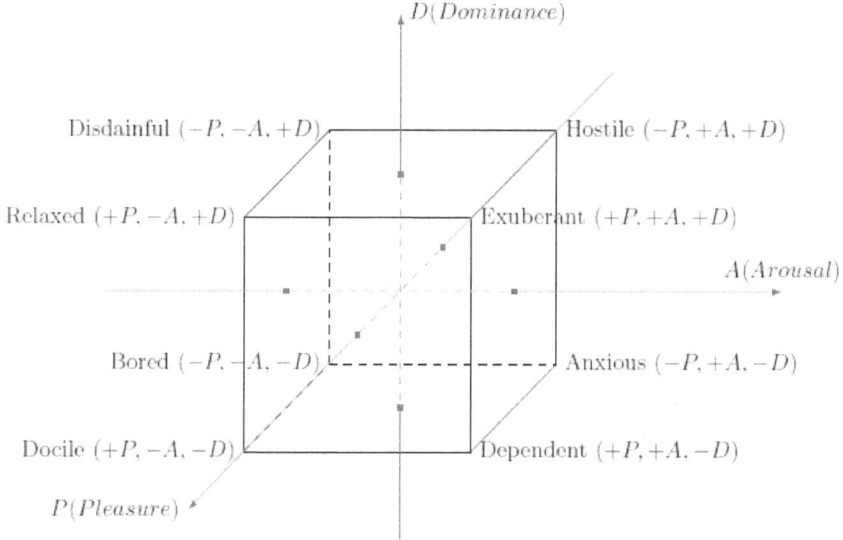

Fig. 4.1 The Pleasure-Arousal-Dominance (PAD) model's space.

Hereafter, we discuss how the PAD model can be used in the RGT to emotionally enrich the interactions between people and humans and robots.

4.4 Bridging the PAD and the RGT

The most important issue to fuse RGT inference and PAD Emotional Scale is to build the bridge between two approaches.

By definition, PAD model is spanned by three dimensions. The value of each component continuously ranges from -1 to 1. The notation in the PAD model space presented in Fig. 4.1 are as follows:

1) pair +P vs -P corresponds to *Pleasure* (positive pole: value 1) vs *Displeasure* (negative pole: value -1);

2) pair +A vs -A corresponds to *Arousal* (positive pole: value 1) vs *Non-arousal* (negative pole: value -1); and

3) pair +D vs -D corresponds to *Dominance* (positive pole: value 1) vs *Submissiveness* (negative pole: value -1).

Applications of the Reflexive Game Theory: Advanced Topics

According to Mehrabian (1996), "pleasure vs. displeasure" distinguishes the positive vs. negative emotional states, "arousal vs non-arousal" refers to combination of physical activity and mental alertness, and "dominance vs. submissiveness" is defined in terms of control vs. lack of control.

ThoPAD model operates with continuous values, i.e., for example, emotional state *Curious* is coded as (0.22, 0.62, -0.01).

Mehrabian (1996) defines eight basic states, which are all possible combinations of high vs. low pleasure (+P vs. -P), high vs low arousal (+A vs. -A), and high vs. low dominance (+D vs. -D).

In other words, there eight basic states are all possible combinations of the poles. In total, there are six poles - two for each of three scales. Therefore, there are $2^3 = 8$ possible combinations of poles.

These states (combinations) are considered as extreme states and can be referred as approximations of the intermediate states. For instance, intermediate state "*Angry*" (-0.51, 0.59, 0.25) can be approximated by extreme state "*Hostile*" *(-P,+A,+D)*. Therefore, only eight extreme states can be used to describe the entire variety of intermediate emotional states.

The combinations of poles in the PAD space can be presented as 3-dimensional binary vectors, where 1 corresponds to positive pole (1 in PAD notation) and 0 refers to negative pole (-1 in PAD notation).

The notations 1 and 0 for positive ("Good"/"good") and negative ("Evil"/ "bad") poles, respectively, has been used by Lefebvre (1982, 2001) and are common for entire *Reflexive Psychology*.

Therefore, emotional states, which are combinations of different poles of various dimensions, can be represented as follows: for instance, emotional state *Hostile (-P,+A,+D)* or (-1,1,1) can be substituted for {0,1,1}, while *Relaxed (+P,-A,+D)* or (1,-1,1) can be represented as {1,0,1} in the rescaled coordinates.

The complete set of emotional states represented as binary vectors is

Docile (+P,-A,-D) is coded as {1,0,0};
Anxious (-P,+A,-D) is coded as {0,1,0};
Disdainful (-P,-A,+D) is coded as {0,0,1};
Hostile (-P,+A,+D) is coded as {0,1,1};
Dependent (+P,+A,-D) is coded as {1,1,0};
Relaxed (+P,-A,+D) is coded as {1,0,1};
Exuberant (+P,+A,+D) is coded as {1,1,1};
Bored (-P,-A,-D) is coded as {0,0,0}.

Among eight basic states, there are three special emotional states *Docile (+P,-A,-D)* ≡ {1,0,0}, *Anxious (-P,+A,-D)* ≡ {0,1,0} and *Disdainful (-P,-A,+D)* ≡ {0,0,1}. These three states are the basis of the 3d binary space. Thus, any other five emotional states can be considered as disjunction (notations OR/∪/+) of these three basis vectors (emotional states).

For example, emotional state *Dependent (+P,+A,-D)* is disjunction of basis states *Docile (+P,-A,-D)* and *Anxious (-P,+A,-D)*: *Docile* OR *Anxious* = *(+P,-A,-D)* cup *(-P,+A,-D)* = {1,0,0} cup {0,1,0} = {1,1,0} = *(+P,+A,-D)* = *Dependent*.

4.5 Merging the RGT and the PAD

Summarizing we highlight the facts about the RGT and the PAD model. According to personal discussions with Lefebvre, the RGT has been proven to predict human choices in the groups of people and allows to control human behavior by means of particular influences on the target individuals.

Next, we note that the PAD model provides description of how the emotional states of humans can be modeled, meaning that a certain emotional state of a particular person can be changed to the desired one.

Furthermore, it is straightforward to see that the coding of the PAD emotional states and alternatives of Boolean algebra are identical.

Therefore, we assume that it is possible to change the emotional states of the subjects in the groups by making influences as elements of the Boolean algebra. In such a case, vector {1,0,0}, for example, plays a role of influence towards emotional state Docile.

Besides, we have distinguished three basis emotional states *Docile* ({1,0,0}), *Anxious* ({0,1,0}) and *Disdainful* ({0,0,1}). The interactions (as defined by disjunction and conjunction operations) of these basic emotional states can result in derivative emotional states such as *Dependent, Relaxed*, etc.

Before, considering the example of PAD application in RGT, we note that reflexive function Φ defines state, which subject is going to switch to. This process goes unconsciously.

Another important issue is that often people directly express their emotions in a form of actions. Therefore in such a case, a particular emotional state of a subject can be considered as the influence on the other subjects.

Example 1. Subjects a and b are in alliance relationship. Subject a makes influence *Dependent* {1,1,0}. Subject b makes influence *Relaxed* {1,0,1}. Their resultant influence will be $(a \cdot b) = \{1,1,0\}\{1,0,1\} = \{1,0,0\}$ or *Docile*. Consequently, the influence of the group, including subjects in alliance with each other, on a given subject is considered as conjunction (defining compromise of all the subjects in alliance) of the influences of all the subjects' influences.

Example 2. Subjects a and b are in conflict relationship. Subject a makes influence *Docile* {1,0,0}. Subject b makes influence *Disdainful* {0,0,1}. Their resultant influence will be $(a + b) = \{1,0,0\}+\{0,0,1\} = \{1,0,1\}$ or *Relaxed*.

Next, we consider an example of reflexive interactions controlling emotional states.

4.6 Emotionally Colorful Reflexive Games

Consider a group of four subjects - the director d and his advisors a, b and c. Let advisors a, b and c are in alliance with each other and in conflict with director *d*. The graph of such group is presented in Fig.4.2. This groups is described by polynomial *abc+d*.

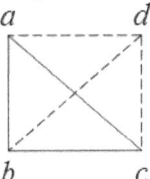

Fig. 4.2 Relationship graph of four subjects *a*, *b*, *c* and *d*.

The canonical form of decision equation for each subject are:

$$a = (bc+d)a + d\overline{a} \quad (4.5)$$
$$b = (ac+d)b + d\overline{b} \quad (4.6)$$
$$c = (ab+d)c + d\overline{c} \quad (4.7)$$
$$d = d + abc\,\overline{d} \quad (4.8)$$

The corresponding solution intervals are

$$(bc+d) \supseteq a \supseteq d \quad (4.9)$$
$$(ac+d) \supseteq b \supseteq d \quad (4.10)$$
$$(ab+d) \supseteq c \supseteq d \quad (4.11)$$
$$1 \supseteq d \supseteq abc \quad (4.12)$$

It is assumed that each subject is in a particular unique emotional state. Let director be in *Exuberant* emotional state. The advisors *a*, *b* and *c* are in *Relaxed* ({1,0,1}), *Docile* ({1,0,0}) and *Anxious* ({0,1,0}) emotional states, respectively.

This variety in emotional states refrains director and his advisors from reaching a fruitful decision. Understanding this emotional situation, director decides to apply reflexive control on emotional level.

Let advisors' influence on all the other subjects coincides with their emotional states, while director is in complete control and can decide, which emotional influence to make on each particular subject.

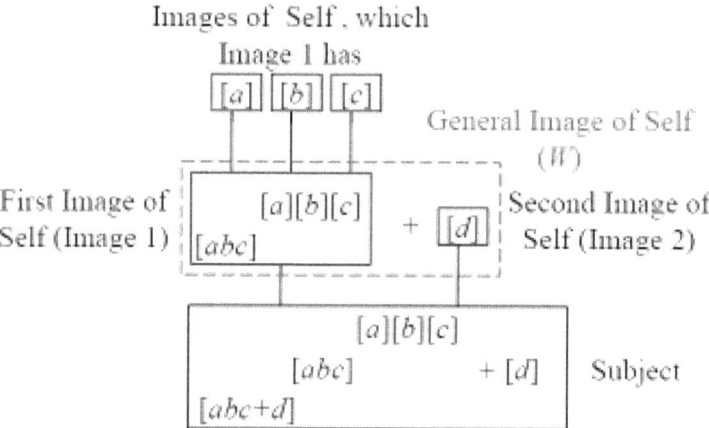

Fig. 4.3 Interpretation of the Diagonal form levels.

Using RGT, we can predict the emotional states of each subject in the group after the *reflexive emotional interaction*:

For subject *a*:
$(\{1,0,0\}\{0,1,0\}+d) \supseteq a \supseteq d \Rightarrow d \supseteq a \supseteq d \Rightarrow a = d;$

For subject *b*:
$(\{1,0,1\}\{0,1,0\}+d) \supseteq b \supseteq d \Rightarrow d \supseteq b \supseteq d \Rightarrow b = d;$

For subject *c*:
$(\{1,0,1\}\{1,0,0\}+d) \supseteq c \supseteq \Rightarrow \{1,0,0\}+d \supseteq c \supseteq d\,;$

For subject *d*:
$1 \supseteq d \supseteq \{1,0,1\}\{1,0,0\}\{0,1,0\} \Rightarrow 1 \supseteq d \supseteq 0 \Rightarrow d=d.$

Therefore, under conditions of such group structure and influences, decisions of advisors *a* and *b* are completely defined by the director's influence.

The entire diagonal form ($P+\overline{W}$) represents the state of the subject. In Fig. 4.3, the diagonal form is marked as Subject. The term $\overline{W}$ is called a general image of the self. On the next level, there are two images - Image 1 ($[abc]^{[a][b][c]}$) and Image 2 ($[d]$). The Images 1 and 2 are images of the self, which general image of the self W has. Finally, the images [a], [b] and [c] are the images of the self, which Image 1 has.

Following this interpretation of the diagonal form (Lefebvre 2009, 2010), we can calculate each emotional state for each image. We analyze the structure of reflexion for advisor c.

His state is ($\{1,0,0\} + d$) or ($Docile$(+P,-A,-D) plus director's influence). The emotional state in the Image 1 is Exuberant ($1 = \{1,1,1\}$), because $[abc]^{[a][b][c]} = [abc] + \overline{[a][b][c]} = 1$.

The Image 2 is [d] that means it is entirely defined by director's influence. Finally, the general image osf the self W is *Exuberant* ($\{1,1,1\}$).

In this simple example, we have illustrated how the emotional states can appear to be the subject for the reflexive control. We call such reflexive control to be *reflexive emotional control*. We have shown how the reflexive emotional control can be successfully implemented by means of the Reflexive Game Theory.

Besides, the RGT allows to unfold the entire sequence of reflexions in the human mind including its emotional aspects.

4.7 Status of a Situation, Multi-stage Decisions and Emotional Reflexive Control

Status of the situation. By analogy with *ethical status* proposed by Lefebvre (1982, 2001), we consider a status of a situation in general. Here, the status of a situation is a probability of the desired outcome, which is to choose a target alternative.

We have already discussed example of the reflexive games of humans and robots in Chapter 2. Here we extend the example about robots baby-sitters with ideas of multi-stage decision making and emotional reflexive control.

Consider that our goal is to make kid a choose alternative $\{\beta\}$. According to eq.(2.2), the choice of kid a is a decision function of three variables b, c and d.

Each variable can take four different values. Therefore, there are $64 = 4 \times 4 \times 4$ different possible inputs (b,c,d) to decision function.

The status of the situation is then the ratio of number of inputs resulting in selection of alternative $\{\beta\}$ to the total number of inputs.

To calculate the status of the situation, we apply the following analysis. Conjunctions cd can take four different values $0=\{\}$, $\{\alpha\}$, $\{\beta\}$ and $1 = \{\alpha,\beta\}$. In general, there are $16 = 4 \times 4$ possible combinations of pairs c and d. Out of these 16 combinations, conjunction of nine, one, three and three combinations result in selection of alternative $0=\{\}$, $\{\alpha\}$, $\{\beta\}$ and $1 = \{\alpha,\beta\}$, respectively.

Each value of conjunction cd results in particular solution intervals for kid a:

cd=0: $b \supseteq a \supseteq 0$;
cd=$\{\alpha\}$: b +$\{\alpha\}$ $\supseteq a \supseteq \{\alpha\}$;
cd=$\{\beta\}$: b + $\{\beta\}$ $\supseteq a \supseteq \{\beta\}$;
cd=1: b + 1 $\supseteq a \supseteq 1 \Rightarrow a = 1$.

Only in the case, when $cd=\{\beta\}$, it is possible that kid a chooses alternative $\{\beta\}$, if robot b makes influence either 0 or $\{\beta\}$. Therefore the status of the situation "kid a chooses $\{\beta\}$" is $2/64 = 1/32$ approx 0.03.

Since we cannot control the influence of kid c on kid a, nearly in 97% of the cases, the situation will NOT result in selection of target alternative.

On the other hand, if we could control subject's influence on other subjects, it could reduce the number of desired outcomes.

Multi-stage decisions. The concept of multi-stage decision making is based on the idea that the mutual influences in the group are not taken for granted, but they are a result of previous decision making session. In such a case, the decision making process consists of two stages. During the first stage (session), decision about mutual influences is made, while the second stage is the final session, when the target decision is made.

In general, the target decision is a function of relationships in the group, mutual influences, etc. On the other hand, these parameters could be themselves the subjects for reflexive control. Therefore, the decision making process of a single target decision unfolds into the sequence of decision making session, when each parameter of the *final decision making session* is a subject of consideration. The concept of multi-stage decision making processes in the RGT has been introduced in Chapter 3.

In this case, our purpose is to achieve the situation when conjunction cd result in $\{\beta\}$:

$$cd = \{\beta\} \qquad (4.13)$$

Since, subject d is one of the robots, we can control value of variable d and should find the appropriate values of variable c.

To do so, we need to solve eq. (4.13). According to Lemma 2 in Chapter 1, the solution of eq. (4.13) regarding variable c is given by the interval (4.14):

$$d + \{\beta\} \supseteq a \supseteq \{\beta\} \qquad (4.14)$$

The solution exist only if variable b takes values 0 and $\{\beta\}$:
$c = \{\beta\}$ for $d = 1$, and
$1 \supseteq c \supseteq \{\beta\}$ for $d = \{\beta\}$.

Therefore, if we could set the kid c influence to $\{\beta\}$ and robot d applies influence to 1, the conjunction cd results in $\{\beta\}$.

Next we consider the two-stage decision making. On the first stage, we discuss conditions, when robots can make kid *c* make a particular influence on kid *a*.

We analyze the sub-group of the group, consisting of kid *c*, and robots *b* and *d* (Fig. 4.4 a)). The purpose of analysis is to find which joint influences robots should make on kid *c* in order to set his influence on kid *a* to $\{\beta\}$.

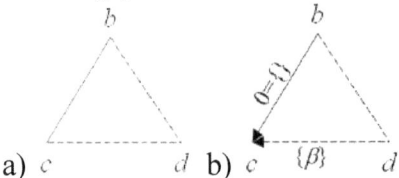

Fig. 4.4 Analysis of sub-group of two robots and kid *c*. a)Relationship graph of robots *b* and *d* and kid *c*. b) Influences of robots *b* and *d* on kid *c* in order to form his influence on kid *a*.

The decision of kid *c* in this sub-group is defined by solution interval:

$$b+d \supseteq c \supseteq d \quad (4.15)$$

To find what *reflexive control (joint influences)* should be applied to kid *c*, the Inverse task should be solved (Chapter 1). The Inverse task is given by the following system of equations:

$$\begin{cases} b+d = \{\beta\} \\ d = \{\beta\} \end{cases} \quad (4.16)$$

System (4.16) results in eq. (4.17):

$$b + \{\beta\} = \{\beta\} \quad (4.17)$$

The solution of eq. (4.17) is given by the interval (4.18):

$$\{\beta\} \supseteq b \supseteq 0 \qquad (4.18)$$

Therefore, in order to set influence of kid c on kid a to $\{\beta\}$, robot d should influence $\{\beta\}$ on him, and robot b should make influence from the interval $\{\beta\} \supseteq b \supseteq 0$.

The sample joint influence of robots on kid c is presented in Fig. 4.4 b).

This guarantees that kid c will make influence $\{\beta\}$ on kid a. Therefore, decision interval for kid a is

$$b+\{\beta\}d \supseteq a \supseteq \{\beta\}d \qquad (4.19)$$

In such a case, there are $16 = 4 \times 4$ possible input (b,d). Out of them, there are four inputs $\{0, \{\beta\}\}$, $\{0,1\}$, $\{\{\beta\},\{\beta\}\}$ and $\{\{\beta\},1\}$ resulting in a situation, when kid a selects alternative $\{\beta\}$ ($a = \{\beta\}$).

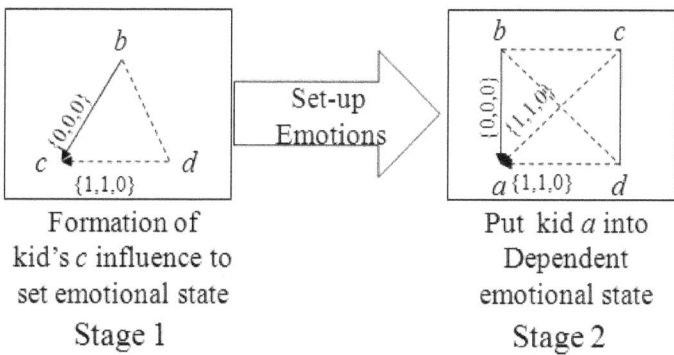

Formation of kid's c influence to set emotional state
Stage 1

Put kid a into Dependent emotional state
Stage 2

Fig. 4.5 Two-stage decision making process with influence pre-set.

In other words, if we now run the process of decision making in the original group, then kid a will choose alternative $\{\beta\}$ with probability 0.25. Therefore, the status of the situation "kid a chooses $\{\beta\}$" increases from ≈ 0.03 to 0.25.

The formation of influence is the first (preliminary) state of the two-stage decision making process. Thus, inclusion of the preliminary stage allows to increase the status of the situation more than eight times (0.25/0.03 ≈ 8.33). The two-stage decision making process is presented in Fig. 4.5.

Nevertheless, there are other methods to increase the status of the situation.

Emotional Reflexive Control. Among the 16 inputs (b,d) considered previously, there are two inputs, which result in the decision interval $\{\beta\} \supseteq a \supseteq 0$. The process of selection of a certain alternative from the decision interval is beyond the scope of the classic Reflexive Game Theory. Therefore, we proposed a method how to select a single alternative on the basis of the emotional state.

We have modeled emotional states as independent instances. Here, we suggest that emotional component is intrinsic for any action, i.e., any alternative of Boolean algebra has two aspects:

1) a meaning of the alternative; and
2) a corresponding emotional state.

In the case with kids, each alternative is characterized from the point of view of physical activity. Since Arousal component in the PAD model is associated with the physical arousal, a particular value of emotional *Arousal* can be associated with each alternative.

The alternative $\{\beta\}$ means "to play with a ball". This implies high value of Arousal: *Arousal* = +*A* or *Arousal*=1, implying that a certain kid is active.

On the other hand, alternative 0={} means to take rest. Therefore, the associated value of *Arousal* is low: *Arousal* = -*A* or *Arousal* = 0, implying that a certain kid is passive.

Since each alternative has corresponding emotional components, we suggest emotional components can be used for a choice of a single alternative from the decision interval.

The choice function can be represented as two "IF-THEN" rules (4.20):

$$Choice = f(Arousal) = \begin{cases} if\ Arousal = 1, then\ \{\beta\} \\ if\ Arousal = 0, then\ 0 \end{cases} \quad (4.20)$$

Next we consider the multi-stage decision making processes including emotional reflexive control.

In order to uniquely define the kid's choice, robot should make influence to shift kid a into a particular emotional state.

Consider robots want the kids "to play with a ball". Out of eight emotional states, there are four states with positive Arousal (+A). The robots decide to move the kids to the *Dependent* state, in which they are completely pleased (+P), but do not consider that situation is under their control (-D). The *Dependent* state is coded as $\{1,1,0\}$.

Because the decision equation for kid a is eq. (2.2), the Inverse task is given by the following system of equations:

$$\begin{cases} b + cd = \{1,1,0\} \\ cd = \{1,1,0\} \end{cases} \quad (4.21)$$

First we solve the second equation regarding variable c. According to Lemma 2 in Chapter 1, its solution is given by the interval:

$$\{1,1,0\}d + \{0,0,1\}\overline{d} \supseteq c \supseteq \{1,1,0\} \quad (4.22)$$

Next we run variable d through all the possible values to obtain the pairs (c,d), satisfying this equation:

$d = 0 = \{0,0,0\}$:
$\{1,1,0\}0 + \{0,0,1\}1 \supseteq c \supseteq \{1,1,0\} \Rightarrow$
$\{0,0,1\} \supseteq c \supseteq \{1,1,0\} \Rightarrow$ no solution;

$d = \{1,0,0\}$:
$\{1,1,0\}\{1,0,0\} + \{0,0,1\}\{0,1,1\} \supseteq c \supseteq \{1,1,0\} \Rightarrow$
$\{1,0,1\} \supseteq c \supseteq \{1,1,0\} \Rightarrow$ no solution;

$d = \{0,1,0\}$:
$\{1,1,0\}\{0,1,0\} + \{0,0,1\}\{1,0,1\} \supseteq c \supseteq \{1,1,0\} \Rightarrow$
$\{0,1,1\} \supseteq c \supseteq \{1,1,0\} \Rightarrow$ no solution;

$d = \{0,0,1\}$:
$\{1,1,0\}\{0,0,1\} + \{0,0,1\}\{1,1,0\} \supseteq c \supseteq \{1,1,0\} \Rightarrow$
$0 \supseteq c \supseteq \{1,1,0\} \Rightarrow$ no solution;

$d = \{0,1,1\}$:
$\{1,1,0\}\{0,1,1\} + \{0,0,1\}\{1,0,0\} \supseteq c \supseteq \{1,1,0\} \Rightarrow$
$\{0,1,0\} \supseteq c \supseteq \{1,1,0\} \Rightarrow$ no solution;

$d = \{1,1,0\}$:
$\{1,1,0\}\{1,1,0\} + \{0,0,1\}\{0,0,1\} \supseteq c \supseteq \{1,1,0\} \Rightarrow$
$1 \supseteq c \supseteq \{1,1,0\} \Rightarrow$
there are two solutions $(1,\{1,1,0\})$ and $(\{1,1,0\}, \{1,1,0\})$;

$d = \{1,0,1\}$:
$\{1,1,0\}\{1,0,1\} + \{0,0,1\}\{0,1,0\} \supseteq c \supseteq \{1,1,0\} \Rightarrow$
$\{1,0,0\} \supseteq c \supseteq \{1,1,0\} \Rightarrow$ no solution;

$d = 1 = \{1,1,1\}$:
$\{1,1,0\}1 + \{0,0,1\}0 \supseteq c \supseteq \{1,1,0\} \Rightarrow$
$\{1,1,0\} \supseteq c \supseteq \{1,1,0\} \Rightarrow$
there is only one solutions $(\{1,1,0\},1)$.

Therefore, there are three (c,d) solutions $(1,\{1,1,0\})$, $(\{1,1,0\}, \{1,1,0\})$ and $(\{1,1,0\}, 1)$ in total.

Next, we solve the first equation of the Inverse task. Since $cd = \{1,1,0\}$, the first equation transforms into eq. (4.23):

$$d + \{1,1,0\} = \{1,1,0\} \qquad (4.23)$$

According to Lemma 1 in Chapter 1, its solution is given by the interval:

$$\{1,1,0\} \supseteq d \supseteq 0 \qquad (4.24)$$

Thus, in order to put kid a into *Dependent* ($\{1,1,0\}$) emotional state robot d should make either influence $\{1,1,0\}$ or influence $0 = \{0,0,0\}$.

On the other hand, robot b should make either influence $\{1,1,1\}$ or $\{1,1,0\}$.

Finally, the kid c should make either influence $\{1,1,0\}$ or 1.

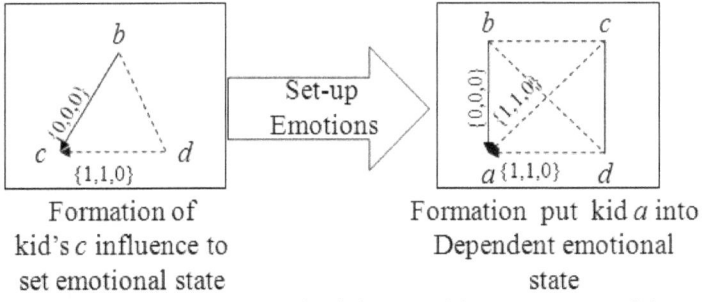

Formation of kid's c influence to set emotional state

Formation put kid a into Dependent emotional state

Fig. 4.6 Two-stage decision making process with emotional state pre-set.

Two robots can communicate with each other and have the same purpose, while kid c is absolutely free to make any influence on kid a. Therefore, to reassure that kid c will shift into required emotional state, robots b and d should make kid c select only influences $\{1,1,0\}$ or 1.

Thus, in order to put kid *a* into appropriate emotional state, robots have first to influence on kid *c* in such way that kid *c* will make a particular influence on kid *a*.

Only after that, robots can put kid *c* into *Dependent* emotional state. This is the first stage of the two-stage emotional decision making process. During the second stage, kid *a* is put into *Dependent* ({1,1,0}) emotional state by efforts of kid *c*, and robots *b* and *d* (Fig. 4.6).

This two-stage emotional control allows to resolve the uncertainty of alternative selection from the interval. Since there are two cases, when solution interval $\{\beta\} \supseteq a \supseteq 0$ occurs (see previous section), then the control of choice allows to raise the status of the situation to value 0.375 (4/16 + 2/16). Therefore, multi-stage decision making involving emotional reflexive control allows to increase the status of the situation.

Overall multi-stage decision making process. The entire multi-stage decision making process is presented in Fig. 4.7.

Stages 1 and 2 are dedicated to setting the required emotional state of kid *a*.

On Stage 1, the robots make influence on kid *c* that he will make influence {1,1,0} on kid *a*. Then on Stage 2, robots *b* and *d* make influences {0,0,0} and {1,1,0}, respectively. Kid *c* makes influence {1,1,0}. Therefore, kid *a* changes emotional state to *Dependent* {1,1,0}. Thus, the required emotional state of kid *a* is achieved.

Next, on Stage 3, robot *b* make influence 0={} and robot *d* makes influence {β} on kid *c* that results in that kid *c* makes influence {β} on kid *a* during the final decision.

On Stage 4, kid *c* make influence {β}, robot *b* makes influence 0={}, and robot *d* makes influence 1={α,β} on kid *a*. This results in that decision of kid *a* is defined by the interval $\{\beta\} \supseteq a \supseteq 0$.

Therefore, kid *a* chooses either active alternative {β} ("to play with a ball") or passive alternative 0={} ("to take a rest"). This concludes the RGT inference.

Finally, on Stage 5 kid *a* will choose alternative {β} ("to play with a ball") according to (4.20), because kid *a* is in *Dependent* ({1,1,0}) emotional state with high physical activity level (*A*=1).

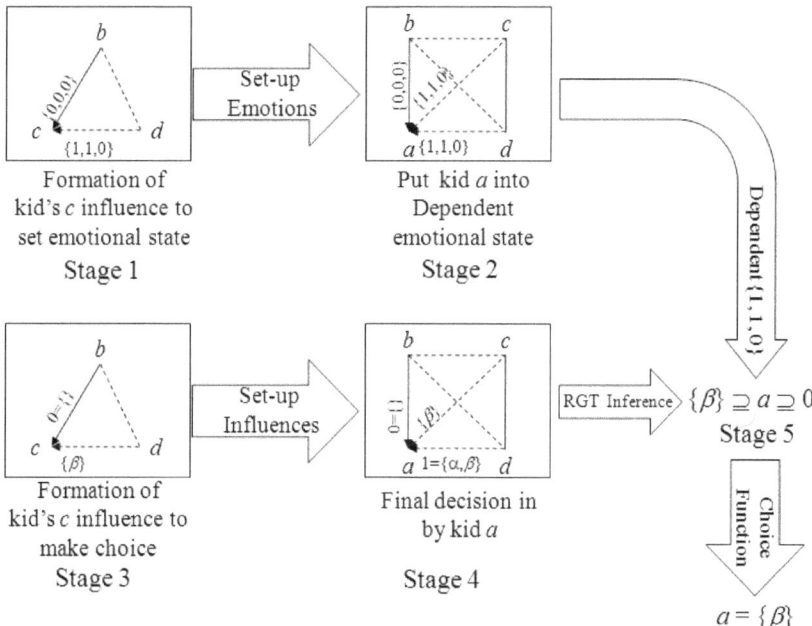

Fig. 4.7 Multi-stage decision making process including emotional reflexive control.

In this example, we have shown how by means of emotional reflexive control more precise reflexive control can be realized. Application of multi-stage decision making and emotional reflexive control helps to increase value of status about 12.5 times from initial value of 0.03 to 0.375.

4.8 Concluding remarks

In this chapter, we have introduced the bridge between the PAD model of emotional states and the RGT calculus by coding PAD's emotional states in terms of the RGT's Boolean algebra of alternatives. We also illustrated how the emotional states can be the subject of reflexive control and can be successfully managed.

This is possible, because the variables in the PAD model, namely, *Pleasure, Arousal, Dominance*, are proved to be necessary and sufficient variables to control human emotions (Russel and Mehrabian, 1977). In other words, necessity means that if emotional state is changed the value of all three variables is changed accordingly. On the other hand, if the values of the variables are changed, the emotional state is also changed.

We have illustrated how to merge the RGT calculus with the PAD model by proper coding emotional states. Finally, we have provided a simple explanatory example of how reflexive emotional control can be applied in action.

Furthermore, we have not only illustrated how to apply the RGT to control emotions of subjects, but also have uncovered the entire cascade of human reflexions as a sequence of *subconscious reflexion*, which allows to trace emotional reflexion of each reflexive image.

This provides us with unique ability to unfold the sophisticated structure of reflexive decision making process, involving emotions.

To date, there has been no approach, capable of doing such thing. This chapter introduces the brand new approach to modeling of human emotional behavior. We call this new breed of the RGT application to be *emotional Reflexive Games (eRG)*.

At present the RGT fused with the PAD model is the unique approach allowing to explore the entire diversity of human emotional reflexion and model reflexive interaction taming emotions. This fusion is automatically formalized in the algorithms and can be easily applied further for developing emotional robots.

We assume, that the proposed mechanism has no heavy negative impact on human psychological state, robots should be enabled to deal with such approach in order to provide human subjects with stress free psychological friendly environments for decision making.

Next, we discuss more complex question regarding emotion and reveal the facts of why the Reflexive Game Theory is the natural approach to model emotions.

We start from trying to answer the question - what are the emotions themselves?

According to Ekman (2007), emotions should be regarded as highly automatics processing algorithms responsible for our everyday life survival. Being entire unconscious (automatic) regulators of the human body's physiology, emotions bring the instantaneous solutions of how to act in front of rapidly approaching possible threat.

Therefore, emotions are fast processors of the information coming apart from environment. Usually, the emotions are characterized by some physiological patterns of body activity on the one hand, and external expression by face mimic or gestures on the other hand.

Furthermore, the reproduction by the self of physical part of emotional pattern, i.e., just making an angry face can elicit the anger as emotional state in the self (Ekman and Davidson, 1993; Levenson et al., 1990). This enables imitation of other's feelings by self-mimicking.

This imitation is interesting from the point of view of reflexive processes. Lefebvre (1977) writes in his book "Structure of Awareness":

"Reflexion in traditional philosophical psychological sense is ability to place one's mind into external position of `observer', `researcher' or 'controller' of one's own body, actions and thoughts. We extend such understanding of reflexion and will consider that reflexion is ability to place one's mind into external position to another 'character', his actions and thoughts.".

From this point of view, the self-mimicking is reflexive process. Thus, it is possible to elicit and understand the emotions of

others by reflexion. Ekman and Davidson (1993)(p. 1210) suggested that this is possible due to
"... direct connections between motor cortex and hypothalamus that translates between emotion-prototypic expression in the face and the emotion-specific pattering in the autonomic nervous system".

On the other hand, the reflexive processes are at the core of the RGT. Therefore the RGT is natural approach to model emotions.

Continuing the line of logic that emotions can be elicited by self reproducing patterns of motor activity accompanying expression of emotions, we suggest the not only facial expressions, but also other motor patterns such as gestures can elicit the corresponding emotion.

Consequently wide variety of motor actions can cause some emotional states to emerge. On the other hands, other actions, not related to the physical activity, can result in particular emotions. Therefore, it is possible to control human emotions not directly, but making people to perform particular actions. From this perspective, PAD model can serve as a measuring device of how actions influence emotions.

However, since the emotions are automatic processing algorithms, which take their input apart from environment, the environment itself is natural medium to influence on emotions. This idea is often used in marketing by providing a customer with delightful emotional state using interior of a shop (Yani-de-Soriano and Foxall, 2006).

Therefore, the continuously on-going decision making process of selecting and implementing actions, is always accompanied by emotional processes, which characterize each action from the emotional perspective.

The fusion of the RGT and the PAD model provides scale free emotional reflexive control framework. Using this framework, it is possible to model different levels of society, which includes not only human beings, but also robotic agents.

Chapter 5: Socializing Autonomous Units

5.1 Introduction

In previous chapters, we constantly use notion of mutual influences between subject. The influences have been considered to be given without questioning how influences are transmitted between subjects.

In this chapter, we consider an example of how communications system between autonomous (robotic) subjects can be established on the basis of frequency modulation. We use the abstract autonomous units, which are capable of communicating with each other in the frequency domain. Therefore, these units can distinguish between several frequencies. Each frequency can be used as a carrying frequency to transmit information between the autonomous units.

To be capable of distinguishing between various frequencies the autonomous units are supplied with frequency selective devices - resonators. There are many possible implementations of the resonators. In this chapter, we use the "Resonate-and-Fire" linear neural model proposed by Izhikevich (2001). The choice of this model is justified by its simplicity of implementation and computations.

5.2 Resonate-and-Fire neurons: a Brief Overview

The original linear model of resonate-and-fire neuron proposed by Izhikevich (2001) is described by the system of two *differential equations*:

$$\begin{cases} \dfrac{dx}{dt} = k\omega - \omega y \\ \dfrac{dy}{dt} = \omega x + ky \end{cases} \quad (5.1)$$

where x is current-like or recovery variable, and y is voltage-like or action potential variable, in terms of neuroscience. Both variables x and y are functions of time: $x = x(t)$ and $y = y(t)$. Notations dx/dt and dy/dt mean time derivatives. Parameter ω is the eigen-frequency, which is preferred or resonant frequency of the system, and represents frequency of sub-threshold oscillations; parameter k is analog of damping factor in the linear damped oscillator. The value of parameter k is set to -0.1 throughout our simulations.

For the purpose of numerical integration the system (5.1) can be transformed into the form (5.2):

$$\frac{dz}{dt} = (k + i\omega)z \qquad (5.2)$$

Then variables x and y are real and imaginary parts of complex variable z, respectively. Using eq.(5.2) and first-order *Euler method*, we straightforwardly obtain *difference equation* (5.3) from differential equation (5.2):

$$z(t+\tau) = z(\tau) + \tau(k + i\omega)z \qquad (5.3)$$

where τ is a small time step. We set τ to 0.005 throughout the simulations.

Iterating difference equation (5.3) with $z(0) = z_0$, we can approximate the analytical solution of eq.(5.2). Now to obtain value of voltage variable $y(t)$ at time t, we only need to take imaginary part of $z(k)$. The real neurons produce a spike, once value of $y(t)$ equal to or exceeds some preset threshold. However, this feature is not provided by the linear model. Therefore we modify the original model by adding the 'spiking' condition:

$$\begin{cases} \text{if } y(t) \geq threshold \quad y(t) = 1.5 \\ y(t+\tau) = 0.1 \end{cases} \qquad (5.4)$$

We set threshold value to 1.0 throughout the simulations. Therefore, the ultimate model of resonate-and-fire neuron is described by the system (5.5)

$$\begin{cases} z(t+\tau) = z(t) + \tau(k+i\omega)z(t) \\ y(t) \geq threshold \ \ y(t) = 1.5 \\ y(t+\tau) = 0.1 \end{cases} \quad (5.5)$$

We present the sample dynamics of two resonate-and-fire neurons, described by system (5.5), with different eigen-frequencies $\omega_1 = 3\pi/2$ and $\omega_2 = 4\pi/3$ in Fig. 5.1.

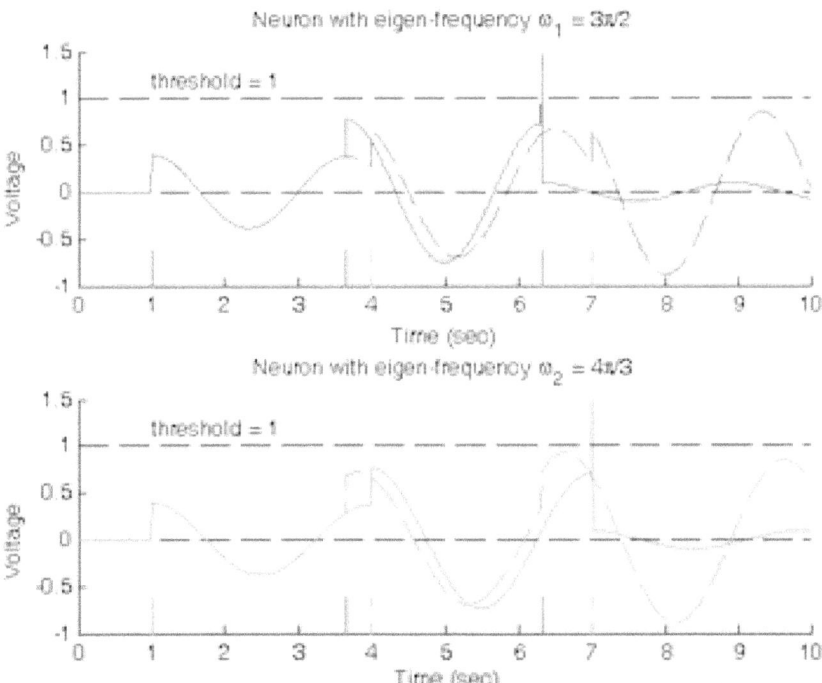

Fig. 5.1 The Resonate-and-Fire neurons.

Applications of the Reflexive Game Theory: Advanced Topics

Top part of Fig 5.1 illustrate dynamics of a neuron with frequency $\omega_1 = 3\pi/2$. Soli line illustrates resonance with the input frequency $\omega_1 = 3\pi/2$, dashed line shows only sub-threshold oscillations meaning that neuron does not respond to the frequency $\omega_2 = 4\pi/3$. In the bottom part of the Fig. 5.1, solid line illustrates resonance with the input frequency $\omega_2 = 4\pi/3$, dashed line shows only sub-threshold oscillations meaning that neuron does not respond to the frequency $\omega_1 = 3\pi/2$.

In Fig.5.1, the solid and dashed vertical lines indicate the equal input pulses of magnitude 0.4. Solid and dashed pulses are provided with frequencies $\omega_1 = 3\pi/2$ and $\omega_2 = 4\pi/3$, respectively. Each series of pulses starts 1 ms after the system onset. Threshold is set to 1.

Thus, we have described the mechanism of frequency selection.

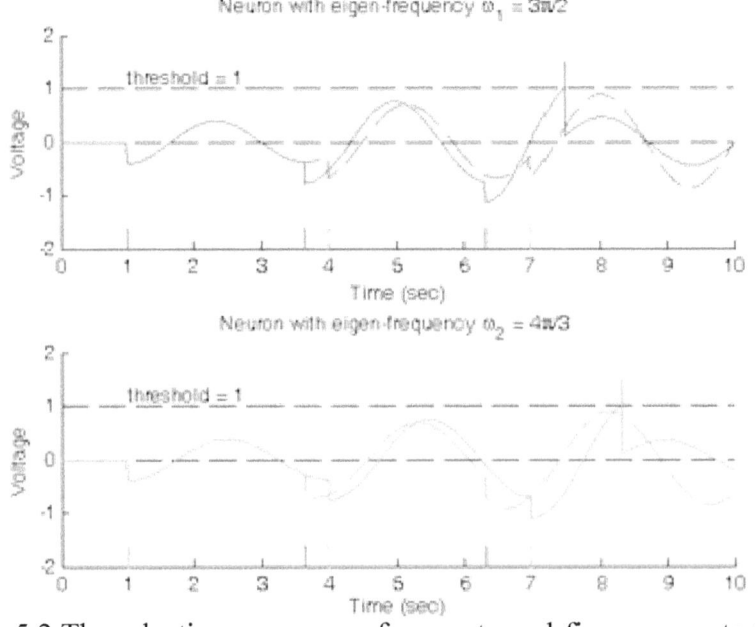

Fig. 5.2 The selective responses of resonate-and-fire neurons to the series of equal inhibitory pulses (magnitude -0.4).

The linear model has other important properties. The inhibitory pulses can also make resonate-and-fire neurons to spike, if the inhibitory (negative) pulses are applied with the eigen-frequency of the neuron (Fig. 5.2).

In Fig.5.2, the top and bottom parts correspond to neurons with eigen-frequences of $\omega_1 = 3\pi/2$ and $\omega_2 = 4\pi/3$, respectively. The solid and dashed vertical lines indicate the equal input pulses of magnitude -0.4. Solid and dashed pulses are provided with frequencies $\omega_1 = 3\pi/2$ and $\omega_2 = 4\pi/3$, respectively.

Next we illustrate that both neurons with eigen-frequencies of $\omega_1 = 3\pi/2$ (Fig. 5.3, top) and $\omega_2 = 4\pi/3$ (Fig. 5.3, bottom) can respond to excitatory series $\{0.1, 0.4, 0.6\}$

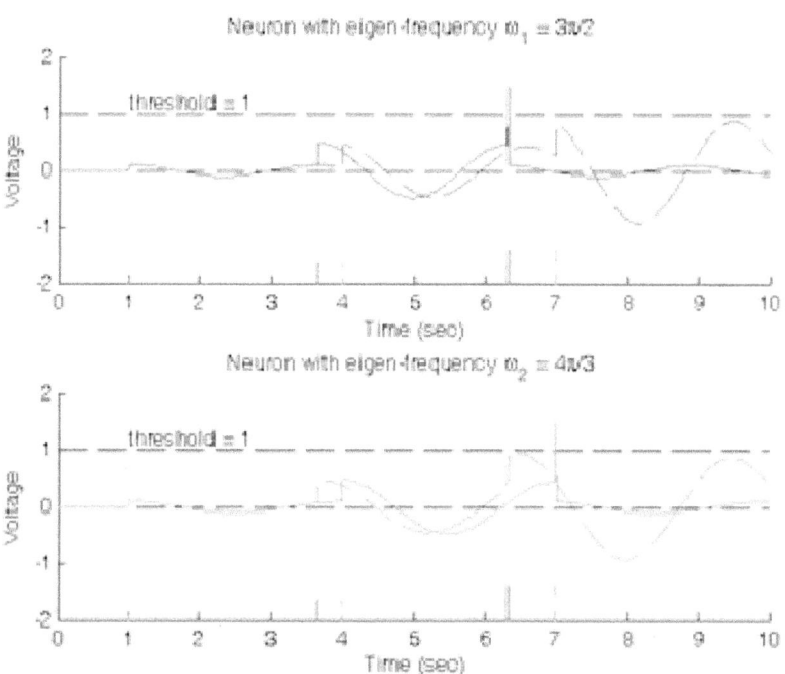

Fig. 5.3 The selective responses of resonate-and-fire neurons to the excitatory series $\{0.1, 0.4, 0.6\}$.

Applications of the Reflexive Game Theory: Advanced Topics

Here we show that both neurons with eigen-frequencies of $\omega_1 = 3\pi/2$ (Fig. 5.4, top) and $\omega_2 = 4\pi/3$ (Fig. 5.4, bottom) can respond to inhibitory series {-0.1, -0.4, -0.6}

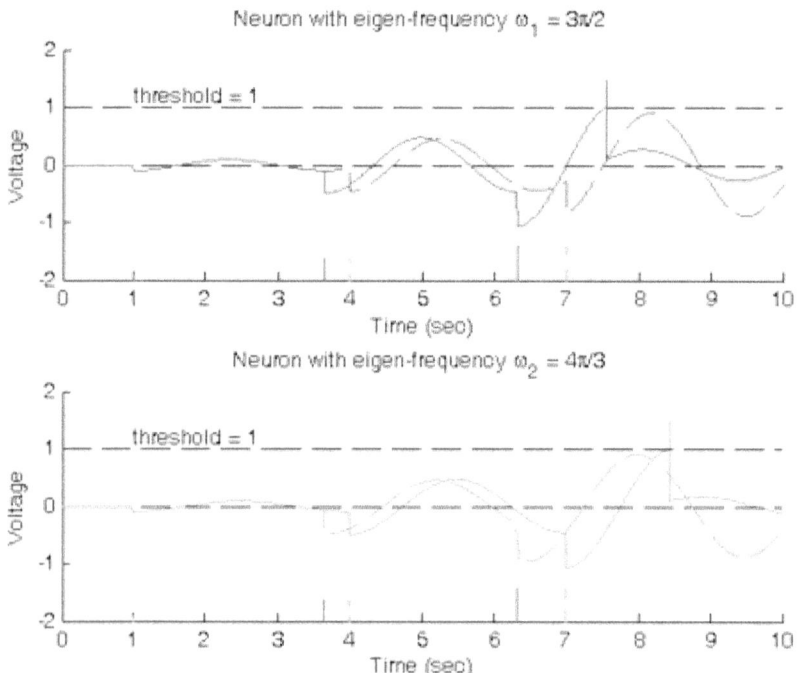

Fig. 5.4 The selective responses of resonate-and-fire neurons to the inhibitory series {-0.1, -0.4, -0.6}.

Since the neurons are selective to a certain frequency, it is possible to transfer signals of several frequencies through the same communication channel as it has been illustrated by Izhikevich (2001). This concludes the description of description of resonate-and-fire neurons. Next we consider the framework to manage the groups of autonomous units.

5.3 Building Communication System for Group to Make Groups of Autonomous Units

Information Coding. Each autonomous unit has several frequency channels. Each channel corresponds to a certain unit in the group. Therefore, the total number of channels equals the total number of units in the group. For each unit, we reserve its unique frequency. Once the information is obtained from this frequency channel, it is considered to be addressed to this unit.

Using the resonate-and-fire neurons, we can transfer different types of information through the network of autonomous units. In fact, we can transmit two types of information coded by

1) the kind of pulses (excitatory vs. inhibitory), and

2) selecting different magnitude of pulses in the series.

These two types of information are enough to model the groups in the Reflexive Game Theory.

We consider a certain frequency to be the unique identifier of the autonomous unit in the group. Next, if the series of pulses contains the excitatory impulses, it is considered that two units are in alliance relationships. They are in conflict otherwise. Finally, we can code a certain alternative of the Boolean Algebra by a certain series of pulses.

This information should be loaded into the autonomous units before their interaction.

Receiving Information in the Group. So far, we understand how a certain unit can receive information. The question remains how autonomous unit can understand where the signal comes from, i.e., which unit sends it?

For this purpose, we reserve a series of equal pulses $\{0.4, 0.4, 0.4\}$. Hereafter, we refer to the series of pulses as code or message.

In particular, we call the code $\{0.4, 0.4, 0.4\}$ to be *identification* (ID)-*code*, if this code is transmitted by a certain unit on its own preferred frequency.

Suggest, we have three units a, b and c. Each unit is characterized by its preferred frequency: $\omega_a = 3\pi/2$, $\omega_b = 4\pi/3$ and $\omega_c = 5\pi/3$.

If autonomous unit a with eigen-frequency ω_a decides to send some code to another unit, it first sends ID-*code* on the frequency ω_a. Therefore, the corresponding neuron spikes in each autonomous units, and units b and c 'understand' that unit a wants to send some code. This can be considered as unit a attracts attention of the other units in the group.

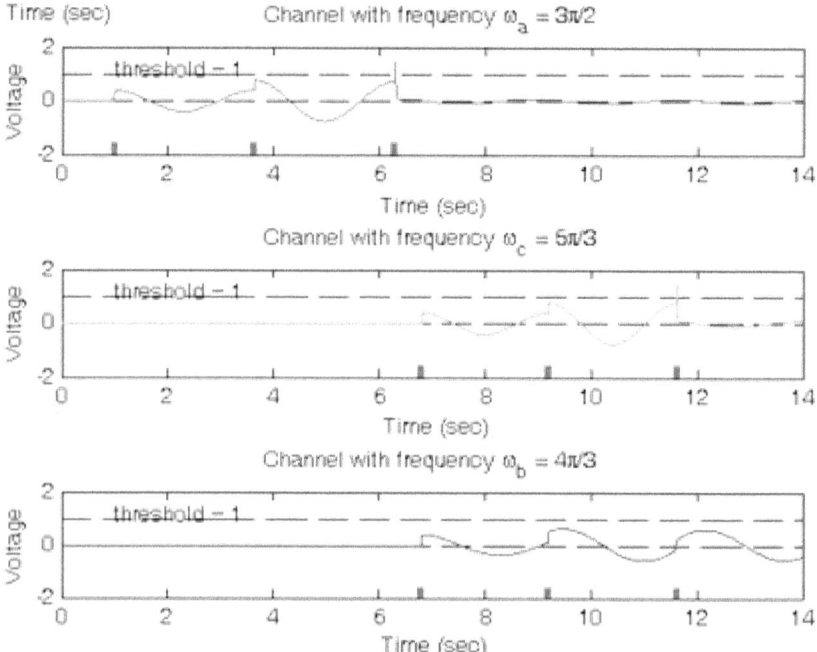

Fig. 5.5 Receiving messages in the network and installing communications.

Then, after a short delay (0.5 sec) after spike on the frequency ω_a, unit a sends a certain code on the frequency ω_x, where x can be either b or c.

As an example, we consider that unit a wants to send its ID-code to unit c. Therefore, unit *a* first sends ID-code {0.4,0.4,0.4} on the frequency ω_a to attract attention of other units: in units *b* and *c* the channels with frequency ω_a show a spike (Fig. 5.5, top).

Then, 0.5 sec after a spike on the frequency ω_a, unit a sends the ID-code on the frequency ω_c: in units *b* and *c* the channels with frequency ω_c show a spike (Fig. 5.5, center).

Since, frequency ω_c is the frequency reserved for unit *c*, unit c receives ID-code from unit *a*. At the same time, channel with frequency ω_b shows no spike (Fig. 5.5, bottom), and unit *b* 'understands' that ID-code is not addressed to it.

This way each unit in the groups can become completely aware about the whole information transmitted between any two units. Therefore, such communication system provides all necessary information for application of the Reflexive Game Theory.

How to Install Relationships in the Group. Now we consider how to install relationship between units. Each unit decides on its own, which type of relationship (conflict or alliance), it wants to install with other units. For example, we consider that the relationships are decided at random, meaning that at the very beginning the units do not have any information about each other, except for the preferred frequencies. Therefore, this condition can be assumes as guessing. The human guessing based on no prior information has been describe from both theoretical and experimental points of views in (Lefebvre 1985, 2006; Tarasenko and Inui, 2009).

In the case of two options, one option (*positive pole*) is chosen with probability $p \approx 0.61$, while another option (*negative pole*) is chosen with probability $1-p \approx 0.39$. The concept of option's polarity has been first introduced by Lefebvre (2006).

We consider alliance relationship to be a positive pole and conflict relationship to be a negative pole.

The alliance and conflict relationships are coded with codes {0.4,0.4,0.4} and {-0.4,-0.4,-0.4}, respectively. We call codes {0.4,0.4,0.4} and {-0.4,-0.4,-0.4} to be alliance and conflict codes, respectively, if they are transmitted NOT on the preferred frequency of the unit, which sends it: 0.5 sec after the ID-code is sent. The alliance and conflict codes are chosen with probabilities 0.61 and 0.39, respectively.

However, to install the relationship, the decisions of both units are needed. In other words, since units choose the type of relationships independently from each other, it is possible that, for example, unit a sends conflict code to unit b, but unit b sends alliance code to unit a.

Therefore, each unit has decided its own relationship, which is different from the one chosen by counterparty. Thus, the codes are different.

We define that the alliance relationship is installed if and only if both units send alliance code to each other, the conflict relationship is installed otherwise. Thus, the relationship between units a and b is conflict.

If we consider codes {0.4,0.4,0.4} and {-0.4,-0.4,-0.4} as logic 1 and 0, respectively, the alliance relationship can be defined as logic conjunction (AND) function, and conflict relationships as disjunction (OR) function.

Table 5.1 Internal state of each unit regarding relationships with others

	a	b	c
a	-	0.81	0.92
b	0.63	-	0.12
c	0.09	0.27	-

Example 1. Let us generate a group of three units with randomly chosen relationships. We use a uniform random number generator with interval (0,1). If the value of random variable exceeds 0.61, unit x generates conflict code {-0.4,-0.4,-0.4} to some other unit, otherwise it generates alliance code {0.4,0.4,0.4}. Sample generated probabilities (Table 5.1). The rows of Table 5.1

contain the decisions about relationships that each unit generated itself, but have not yet transmitted to other units. Therefore, Table 5.1 illustrates internal/inner state of each unit. This internal state is yet not known by other units in the group.

According to Table 5.1, unit a will send conflict code to both units b and c (Fig. 5.6). Unit b will send conflict code to unit a and alliance code to unit c (Fig. 5.7). Unit c will send alliance code to both units b and c (Fig. 5.8).

After the codes have been transmitted from each unit to each unit, the information from Table 5.1 becomes available to each unit in the group. Using the correlation between conflict and alliance codes and logic 1 and 0, we can rewrite Table 5.2.

Table 5.2 Transmitted relationship codes

	a	b	c
a	-	0	0
b	0	-	1
c	1	1	-

Therefore, using information from Table 5.2 together with conjunction and disjunction functions, we obtain the relationships installed between the units: units b and c are in alliance, while unit a is in conflict with both units b and c.

Since, the information about the relationships between units is now known by each unit in the group, we can construct the relationship graph (Fig. 5.9) of the Reflexive Game Theory and obtain the polynomial corresponding to this graph, which is $a+bc$.

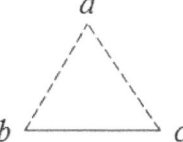

Fig. 5.9 Relationship group of a group of autonomous units.

Applications of the Reflexive Game Theory: Advanced Topics

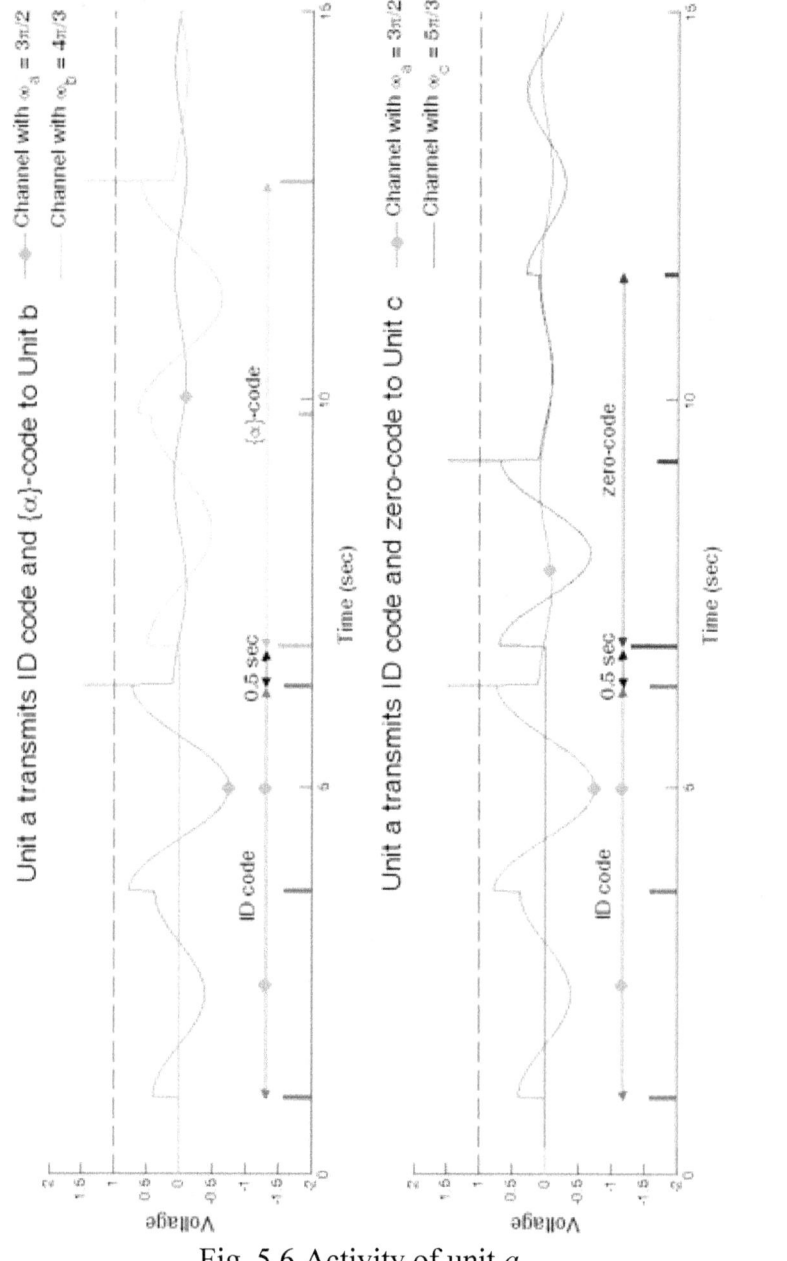

Fig. 5.6 Activity of unit *a*.

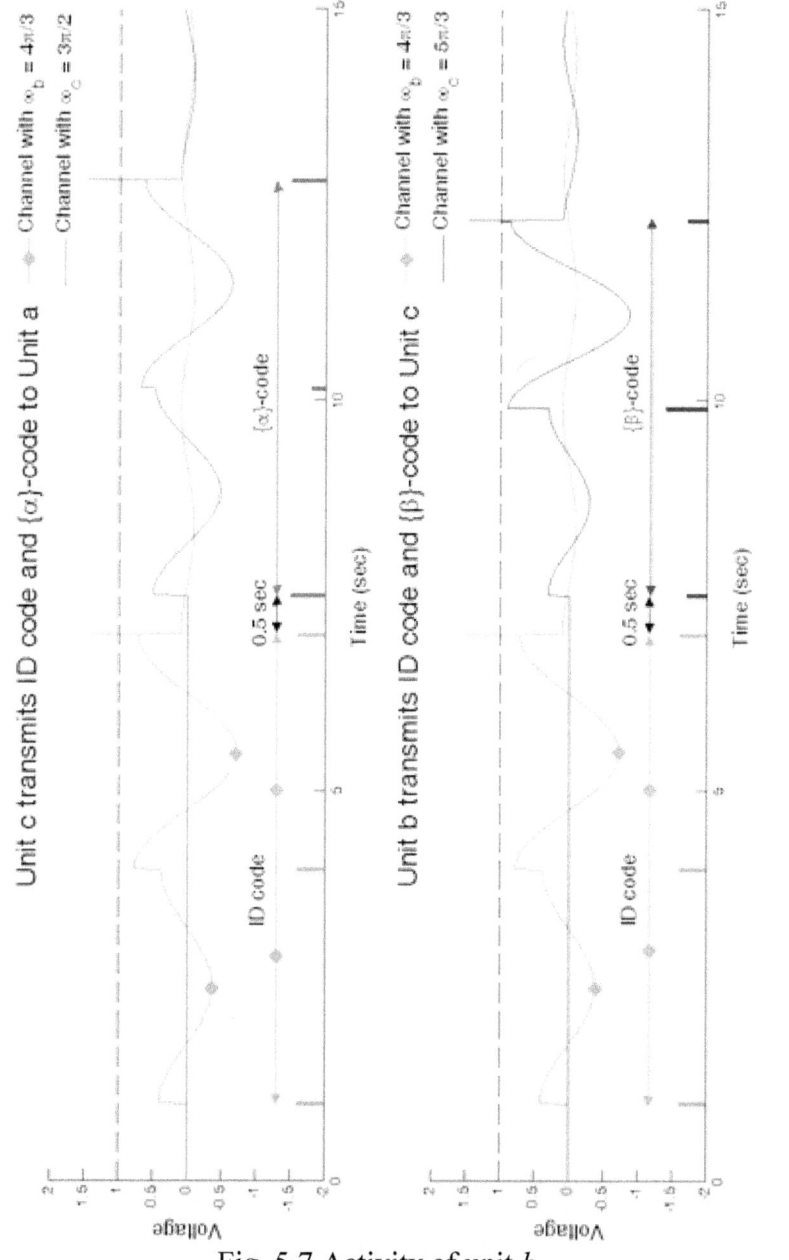

Fig. 5.7 Activity of unit b.

Applications of the Reflexive Game Theory: Advanced Topics

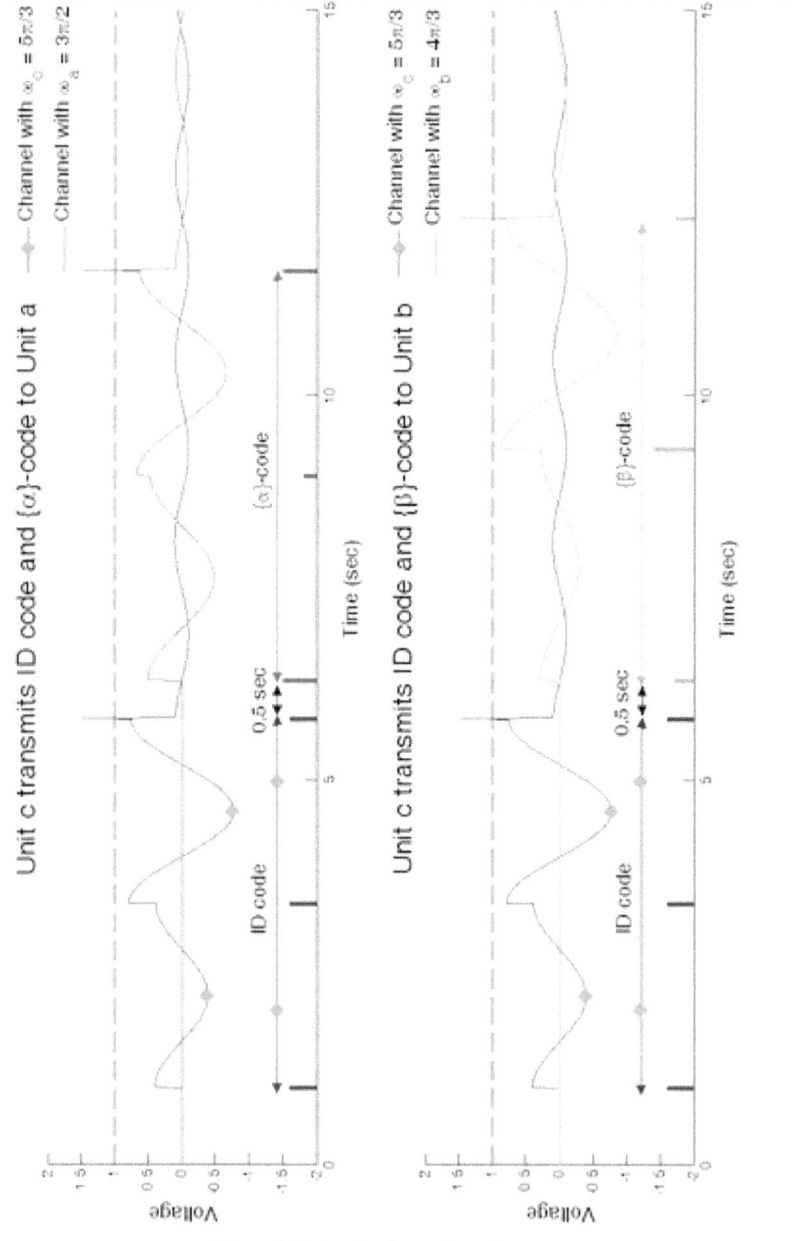

Fig. 5.7 Activity of unit c.

How to Transfer Information about the Influences. In this section, we illustrate how to transmit influences of unit on each other. We use the same approach described in the previous section. The only difference is that instead of the alliance or conflict codes, unit transmits some code associated with a particular alternative.

Example 2. Suggest, we have Boolean algebra of four alternatives: $1 = \{\alpha,\beta\}$, $\{\alpha\}$, $\{\beta\}$ and $0 = \{\}$. We arbitrary assign a certain code to each alternative: 1) code $\{0.2, 0.3, 0.7\}$ corresponds to alternative $1 = \{\alpha,\beta\}$; 2) code $\{0.7, 0.3, 0.2\}$ corresponds to alternative $0=\{\}$; 3) code $\{0.5, 0.2, 0.5\}$ corresponds to alternative $\{\alpha\}$; and 4) code $\{0.3, 0.6, 0.3\}$ corresponds to alternative $\{\beta\}$. To ease of reference, we refer to each code as 'alternative name'-code: code $\{0.7, 0.3, 0.2\}$ is called unit-code, code $\{0.2, 0.3, 0.7\}$ is zero-code, code $\{0.5, 0.2, 0.5\}$ is referred as $\{\alpha\}$-code, and code $\{0.3, 0.6, 0.3\}$ is $\{\beta\}$-code.

We assume that unit *a* makes influences $\{\alpha\}$ and $0=\{\}$ on units *b* and *c*, respectively (Fig. 5.6).

Unit *b* makes influence $\{\alpha\}$ on both units a and c (Fig. 5.7). Unit c makes influences $\{\alpha\}$ and $\{\beta\}$ on units *a* and *b*, respectively (Fig. 5.8).

Table 5.3 Influence matrix

	a	b	c
a	a	$\{\alpha\}$	$\{\}$
b	$\{\alpha\}$	b	$\{\beta\}$
c	$\{\beta\}$	$\{\beta\}$	c

RGT Inference. Therefore, after all the influences have been transmitted, we obtain the influence matrix (Table 5.3). Thus, each unit now has complete information to apply the RGT inference schema based on the decision equations.

The canonical form of decision equation for unit *a* is

$$a = a + bc\overline{a} \qquad (5.6)$$

and the corresponding solution interval is

$$1 \supseteq a \supseteq bc \qquad (5.7)$$

The canonical form of decision equation for unit b is

$$b = (a+c)b + a\overline{b} \qquad (5.8)$$

and the corresponding solution interval is

$$(a+c) \supseteq b \supseteq a \qquad (5.9)$$

The canonical form of decision equation for unit c is

$$c = (a+b)c + a\overline{c} \qquad (5.10)$$

and the corresponding solution interval is

$$(a+b) \supseteq c \supseteq a \qquad (5.11)$$

Under the given influences, the choice of unit a is define by the interval $1 \supseteq a \supseteq \{\}$. The solution interval for unit b turns into equality $b = \{\alpha\}$. The choice of unit a is define by the interval $\{\beta\} \supseteq a \supseteq \{\}$.

5.4 Discussion

In this chapter, we have presented the structure of autonomous units, which allows these units to install communication with each other and create groups. As the basis for communication network, we use resonate-and-fire neurons, which are used as signal receivers. The main feature of resonate-and-fire neurons is their selectivity to a particular frequency, which is eigen-frequency of the neuron. Therefore, it is possible to send different codes through the same network and be sure that each unit understands the message addressed exclusively to it. We do not discuss here physical mechanisms of generating signals.

We have illustrated how to arrange a group of three units by using a communication network. We have also showed how to code different messages such as sender identification and Boolean algebra alternatives.

We have concluded with examples of how a simple group can be arranged based on the information about relationships between units and showed how to transmit the information about influences in the group. Thus, having received the information about the structure of the group and the mutual influences, each autonomous unit can apply algorithms of RGT inferences. Therefore, each unit can make both its own choice and also predict the possible choices of other members of the group.

Therefore, here we have proposed a method of communication between autonomous units and have illustrated how the RGT can be applied to the feasible units by showing how the information needed for the RGT inferences is generated and transmitted between the group members.

Chapter 6: Modeling Social Dynamics

6.1 Social dynamics

The matter of social dynamics is one of the central issues in the social psychology and sociology. Here, we consider the matter of social dynamics from the point of view of newly proposed Reflexive Game Theory.

The material in this chapter is presented as follows. First, we discuss a structure of groups in Reflexive Game Theory and propose a method how to create a certain group structure. Then we analyze how a structure of a given group can be changed from the point of view of the RGT. Finally, we propose approach to model the process of social dynamics using the Markov Chains.

6.2 Formation of relationships

Here we consider a fundamental issue of how a group can emerge from the number of individuals. Under the term "*group emergence*", we understand a result of interaction between several individuals regarding the relationship between them, i.e., individuals communicate with each other to set a pair-wise relationships in a group. When the relationships between all the individuals are set and, thus, a group could be represented in the form of fully connected graph this is the moment, when one can say that a group has emerged.

In this chapter, we assume that the relationships between the individuals could be of either *alliance* or *conflict* types.

In RGT it is considered that conflict is disjunction and alliance is modeled with conjunction operations, respectively. It is assumed the two subjects being in alliance can find the compromise or a common influence, therefore their interaction can be characterized as conjunction of their influences. On the other hand, two conflicting subjects will never reach a compromise and their interaction

Applications of the Reflexive Game Theory: Advanced Topics

should include both influences. Therefore disjunction is used to represent conflict relationship.

In this chapter, we define how the relationships are installed. As in the previous chapter, we postulate that alliance relationship could appear if and only if both subjects consider each other to be friends or allies. If at least one of two individuals does not consider a vis-a-vis as a friend or ally, then conflict relationship appears.

In generally, installation of relationships can be defined with *conjunctions operation*, where 1 means to consider a vis-a-vis to be friend or ally, and 0 means that vis-a-vis is attained as an enemy.

6.3 Representation of the group structure

The RGT uses Boolean algebra of alternatives as the basis for calculus. Each groups in RGT is represented as fully connected group. However, when speaking about the formation of the groups structure, we should define a special way to represent the relationships in the group in terms of the Boolean algebra of alternative.

Here we consider an example of how it can be done. For simplicity we study group of three subjects a, b and c. With respect to the relationships between the subjects, there can be only four different structures:
1) all the subjects are in alliance;
2) all the subjects are in conflict;
3) one subject is in alliance with conflicting subjects;
4) two subjects in alliance are in conflict with the third subject.

In the group of three subjects regarding the location in the graph, each subject has one subject on the left and one subject on the right, if we image to look from the position of a given subject the point in the middle between the other two subjects.

Therefore, the state of each subject regarding relationships with other subjects can be represented in the form of 2d binary vector $\{\alpha,\beta\}$. The first (α) and the second (β) components illustrate the relationships of the given subject with subject to the left and to the right, respectively. Value 1 indicates an intention to make alliance relationship, value 0 implies an intention to make conflict relationship.

The Boolean algebra of alternatives then contains four elements: 1) 1 = {1,1} - alliance with both other subjects; 2) {1,0} - alliance with a subject to the left and conflict with a subject to the right; 3) {0,1} - conflict with a subject to the left and alliance with a subject to the right; 4) {0,0} - conflict with both subjects.

Next we consider the cases when each configuration of the group could emerge.

Alliance group (Type I). Alliance group means that all the subject in the group are in alliance with each other. The alliance group of subjects a, b and c is presented in Fig.6.1a). Emergence of this group is possible if and only if all three subjects choose alternative 1: $a = b = c = \{1,1\} = 1$.

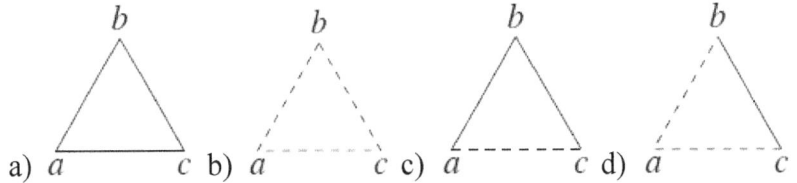

Fig. 6.1 Possible group structures for three individuals.

Table 6.1 Possible combinations of influences to install a group of Type II

a	b	c	a	b	c	a	B	c
{0,0}	{0,0}	{1,1}	{0,0}	{0,1}	{0,1}	{0,1}	{1,1}	{0,0}
{0,0}	{0,0}	{1,0}	{0,0}	{1,1}	{1,0}	{0,1}	{1,0}	{0,0}
{0,0}	{0,0}	{0,1}	{0,0}	{1,1}	{0,0}	{0,1}	{0,1}	{0,0}
{0,0}	{0,0}	{0,0}	{1,1}	{0,0}	{0,0}	{1,0}	{0,0}	{0,0}
{0,0}	{1,0}	{1,0}	{1,1}	{0,0}	{0,1}	{1,0}	{0,0}	{1,0}
{0,0}	{1,0}	{0,0}	{1,1}	{1,0}	{0,0}	{1,0}	{0,0}	{0,1}
{0,0}	{0,1}	{0,0}	{0,1}	{0,0}	{0,0}	{1,0}	{0,0}	{1,1}
{0,0}	{0,1}	{1,0}	{0,1}	{0,0}	{0,1}	{1,0}	{1,0}	{0,0}
{0,0}	{0,1}	{0,1}	{0,1}	{0,1}	{1,0}	{1,0}	{1,0}	{1,0}

Conflict group (type II). Conflict group means that all the subject in the group are in conflict with each other. The conflict group of subjects *a*, *b* and *c* is presented in Fig. 6.1b).

The group structures in terms of elements of Boolean algebra is presented in Table 6.1.

One subject is in alliance with two conflicting subjects (type III). The example of such group of subjects *a*, *b* and *c* is presented in Fig.6.1c). Such group can emerge if, for example, subject *b* wishes to be in alliance in both subjects, and other subjects want to be in alliance with subject *b*, but there is no alliance between them. The possible group structures in term of Boolean are presented in Table 6.2.

In general, the conflict relationship could be between any two subjects in a group, therefore, in total there are 9 (3x3) structures of the groups having one subject in alliance with other two conflicting subjects.

Table 6.2 Possible combinations of influences to install a group of Type III

a	*b*	*c*
{1,0}	{1,1}	{0,1}
{1,0}	{1,1}	{1,1}
{1,1}	{1,1}	{0,1}

Table 6.3 Possible combinations of influences to install a group of Type IV

a	*b*	*C*	*a*	*b*	*c*
{0,0}	{1,0}	{0,1}	{0,1}	{1,1}	{0,1}
{0,0}	{1,1}	{0,1}	{0,1}	{1,0}	{0,1}
{0,0}	{1,0}	{1,1}	{1,0}	{1,0}	{1,1}
{0,0}	{1,1}	{1,1}	{1,0}	{1,0}	{0,1}
{1,1}	{1,0}	{0,1}			

One subject is in conflict with two subjects in alliance (type IV). The example of such group of subjects *a*, *b* and *c* is presented in Fig6.1d). The group structures in terms of elements of Boolean algebra is presented in Table 6.3.

In general, the alliance relationship could be between any two subjects in a group, therefore, in total there are 27 (9 x 3) structures of the groups having one subjects in conflict with other two subjects in alliance.

As a conclusion of this section, we discuss the case when the structure of the group is generated randomly. In total, there are $4^3=64$, where 4 is a number of elements of Boolean algebra, 3 is a number of subjects in a group, possible combinations resulting in structures of the group: 1 combination results in alliance group (Type I), 27 combinations results in conflicting groups (type II), 9 combinations results in groups with one subject in alliance with two conflicting subjects (Type III) and 27 combination results in groups with to subject in alliance and in conflict with another subject (Type IV).

The corresponding probabilities are
P(Type I) = 1/64,
P(Type II) = 27/64,
P(Type III) = 9/64,
P(Type IV) = 27/64.

6.4 Transition between different group structures

In this section, we consider how the group structure could be changed. We have shown in the previous section that there are four different types of the structures.

In total, there are eight different groups. Here we list them in the form of the corresponding polynomials: [*abc*], [*a+b+c*], [*ab+c*], [*ac+b*], [*bc+a*], [*a(b+c)*], [*b(a+c)*] and [*c(a+b)*]. Square brackets here are used to indicate the polynomials.

Next we consider the elements of Boolean algebra as influences, which subjects make on each other, and calculate the transition probabilities from one group structure to another.

After the decision making, each subject will make a particular decision regarding the influences on him from other subjects. All the decisions that subject makes are called a *decision spectrum*. It is possible to calculate the probability of each decision under condition, when all possible combinations of influences on him are applied. The probability distribution for decision spectrum is called a *probability spectrum*. The decisions from decision spectrum can be different from particular alternatives of the Boolean algebra. Therefore we consider another spectrum which contains only alternatives themselves and call this spectrum to be *spectrum of alternative* and its corresponding *probability spectrum of alternative*.

Because the structures of the groups are presented in the form of combinations of alternatives in Tables 6.1~6.3, the probability spectrum of alternative is crucial in understanding transition probabilities from one group to another.

Next we consider the spectra of alternative and probability spectra of alternatives, which each subject has being involved in a particular group structure.

First, we discuss the groups $[abc]$ and $[a+b+c]$.

After the diagonal form folding we obtain a single alternative $1=\{1,1\}$: $[abc]^{[a][b][c]} = [abc] + \overline{[a][b][c]} = 1$ and $[a+b+c]^{[a]+[b]+[c]} = [a+b+c] + \overline{[a]+[b]+[c]} = 1$. Therefore all the subject in the group choose alternative $1=\{1,1\}$, meaning make alliance with other subjects, with probability P equal 1.

Therefore, if the group is $[abc]$, it stays the same with probability $P([abc] \rightarrow [abc])=1$. If the group is $[a+b+c]$, it changes into group $[abc]$ with probability $P([a+b+c] \rightarrow [abc])=1$.

In other words, the spectra of alternatives and probability spectra of alternatives of subjects in these two groups are restricted to the single alternative 1={1,1} with probability 1. Next we consider these spectra in the other groups.

Analysis of a Type III group. Here we consider a single example of the group $[a(b+c)]$. The decision interval for subject a, b and c are presented in Table 5.4.

Table 6.4 Decision intervals for the subjects in a group of Type III.

a	b	c
$b+c \supseteq a \supseteq 1$	$1 \supseteq b \supseteq ac+\bar{a}$	$1 \supseteq c \supseteq ab+\bar{a}$

Table 6.5 All possible realization of the decision intervals in presented in Table 5.4 for Type III group.

b	c	a	a	c	b	a	b	c
{0,0}	{0,0}	NA	{0,0}	{0,0}	b=1	{0,0}	{0,0}	c=1
{0,0}	{0,1}	NA	{0,0}	{0,1}	b=1	{0,0}	{0,1}	c=1
{0,0}	{1,0}	NA	{0,0}	{1,0}	b=1	{0,0}	{1,0}	c=1
{0,0}	{1,1}	a=1	{0,0}	{1,1}	b=1	{0,0}	{1,1}	c=1
{1,0}	{0,0}	NA	{1,0}	{0,0}	$1 \supseteq b \supseteq \{0,1\}$	{1,0}	{0,0}	$1 \supseteq c \supseteq \{0,1\}$
{1,0}	{1,0}	NA	{1,0}	{1,0}	b=1	{1,0}	{1,0}	c=1
{1,0}	{0,1}	NA	{1,0}	{0,1}	$1 \supseteq b \supseteq \{0,1\}$	{1,0}	{0,1}	$1 \supseteq c \supseteq \{0,1\}$
{1,0}	{1,1}	a=1	{1,0}	{1,1}	b=1	{1,0}	{1,1}	c=1
{0,1}	{0,0}	NA	{0,1}	{0,0}	$1 \supseteq b \supseteq \{1,0\}$	{0,1}	{0,0}	$1 \supseteq c \supseteq \{1,0\}$
{0,1}	{1,0}	NA	{0,1}	{1,0}	$1 \supseteq b \supseteq \{1,0\}$	{0,1}	{1,0}	$1 \supseteq c \supseteq \{1,0\}$
{0,1}	{0,1}	NA	{0,1}	{0,1}	b=1	{0,1}	{0,1}	c=1
{0,1}	{1,1}	a=1	{0,1}	{1,1}	b=1	{0,1}	{1,1}	c=1
{1,1}	{0,0}	a=1	{1,1}	{0,0}	$1 \supseteq b \supseteq 0$	{1,1}	{0,0}	$1 \supseteq c \supseteq 0$
{1,1}	{1,0}	a=1	{1,1}	{1,0}	$1 \supseteq b \supseteq \{1,0\}$	{1,1}	{1,0}	$1 \supseteq c \supseteq \{1,0\}$
{1,1}	{0,1}	a=1	{1,1}	{0,1}	$1 \supseteq b \supseteq \{0,1\}$	{1,1}	{0,1}	$1 \supseteq c \supseteq \{0,1\}$
{1,1}	{1,1}	a=1	{1,1}	{1,1}	b=1	{1,1}	{1,1}	c=1

Next for each subject we calculate the possible realizations of decision intervals, which are presented in Table 5.4 . The results are presented in Table 5.5. Regarding Table 5.5, we can calculate probability spectrum for each subject.

For subject a. The decision spectrum of subject *a* contains only one decision 1={1,1}. The probability spectrum is $P(a = \{1,1\})$ = 7/16. In other cases, subject a is in a frustration state (*NA*) and cannot make any decision. Therefore, $P(a$ is in frustration$) = 9/16$.

For subject b. For subject *b* decision spectrum includes four decisions:

$1 \supseteq b \supseteq 0$,
$1 \supseteq b \supseteq \{0,1\}$,
$1 \supseteq b \supseteq \{1,0\}$,
$b=1$.

Next we calculate the probability spectrum for subject b:
$P(1 \supseteq b \supseteq 0) = 1/16$;
$P(1 \supseteq b \supseteq \{0,1\}) = 3/16$;
$P(1 \supseteq b \supseteq \{1,0\}) = 3/16$;
$P(b =1) = 9/16$.

The spectrum of alternatives for subject *b* contains all four alternatives from Boolean algebra. These alternatives can be selected both as a result of decision making and as a result of realization of a single alternative from the interval of alternatives.

If the alternative can be selected from the given interval, we consider that all the alternatives contained by the interval have the same probability to be realized.

Therefore each probability from the probability spectrum of alternatives is a *marginal probability* of all the possible outcome when the given alternative can be selected or realized. We use notation $P_m(\cdot)$ to indicate the marginal probabilities.

The marginal probability that subject *b* chooses alternative 1 $P_m(b=1)$ is computed as follows:

$P_m(b=1) = P(b=1) + P(b=1|1\supseteq b\supseteq 0) + P(b=1|1\supseteq b\supseteq\{0,1\})) + P(b=1|(1\supseteq b\supseteq\{1,0\}))$.

Probability $P(b=1|(1\supseteq b\supseteq 0)$ is the probability that alternative 1 will be realized from the intervals $(1\supseteq b\supseteq 0)$. This is conditional probability of alternative to be selected from the interval $1\supseteq b\supseteq 0$, when decision $1\supseteq b\supseteq 0$ is made.

Therefore, conditional probability

$$P(b=1|1\supseteq b\supseteq 0) = P(1\supseteq b\supseteq 0)\, P(b=1, 1\supseteq b\supseteq 0),$$

where $P(1\supseteq b\supseteq 0)$ is a probability to make decision $1\supseteq b\supseteq 0$, and $P(b=1, 1\supseteq b\supseteq 0)$ is a probability to select alternative 1 from the interval $1\supseteq b\supseteq 0$.

The probability $P(1\supseteq b\supseteq 0)$ is contained in the probability spectrum, and $P(b=1, 1\supseteq b\supseteq 0)$ is 1/4, since all the alternatives within a given interval are equally probable.

Therefore,
$P_m(b=1) =$
$P(b=1) + P(1\supseteq b\supseteq 0)\, P(b=1, 1\supseteq b\supseteq 0) +$
$P(1\supseteq b\supseteq\{0,1\})\, P(b=1, 1\supseteq b\supseteq\{0,1\}) +$
$P(1\supseteq b\supseteq\{1,0\})\, P(b=1, 1\supseteq b\supseteq\{1,0\})$.

Finally,
$P_m(b=1) = 9/16 + 1/16\, P(b=1, 1\supseteq b\supseteq 0) +$
$3/16\, P(b=1, 1\supseteq b\supseteq\{0,1\}) + 3/16\, P(b=1, 1\supseteq b\supseteq\{1,0\})$.

$P_m(b=1) = 9/16 + 1/16 \cdot 1/4 + 3/16 \cdot 1/2 + 3/16 \cdot 1/2 = 49/64$.

The marginal probability $P_m(b=\{1,0\})$ is
$P_m(b=\{1,0\}) = P(b=\{1,0\}|1\supseteq b\supseteq\{1,0\}) + P(b=\{1,0\}|1\supseteq b\supseteq 0)$
since alternative $\{1,0\}$ can be realized if either decision $1\supseteq b\supseteq\{1,0\}$ or $1\supseteq b\supseteq 0$ is made.

Therefore, the marginal probability $P_m(b=\{1,0\})$ is

$P_m(b=\{1,0\}) = P(b=\{1,0\}|1\supseteq b\supseteq\{1,0\}) + P(b=\{1,0\}|1\supseteq b\supseteq 0) =$
$P(1\supseteq b\supseteq\{1,0\})P(b=\{1,0\}, 1\supseteq b\supseteq\{1,0\}) +$
$P(1\supseteq b\supseteq 0) \ P(b=\{1,0\}, 1\supseteq b\supseteq 0)=$
$3/16\cdot1/2 + 1/16\cdot1/4 = 7/64$.

The marginal probability $P_m(b=\{0,1\})$ can be computed similar to $P_m(b=\{1,0\})$ and is also 7/64.

The marginal probability $P_m(b=0)$ is
$P_m(b=0) = P(b=0|(1 \supseteq b \supseteq 0))$
since alternative 0 can be realized only if decision is $1\supseteq b\supseteq 0$.
$P_m(b=0)$ is 1/64.

Now we confirm that marginal probabilities of all four alternatives sum up to 1:
$P_m(b=1) + P_m(b=\{1,0\}) + P_m(b=\{0,1\}) + P_m(b=0) =$
$49/64 + 7/64 + 7/64 + 1/64 = 1$.

For subject c: the probability spectrum and marginal probabilities (probability spectrum of alternatives) are calculated in the same manner as for subject b and are the same.

Analysis of a Type IV group. Here we consider a single example of the group $[ab+c]$. The decision interval for subject *a*, *b* and *c* are presented in Table 6.6.

Table 6.6 Decision intervals for the subjects in a group of Type IV.

a	b	c
$b+c \supseteq a \supseteq c$	$a+c \supseteq b \supseteq c$	$1 \supseteq c \supseteq ab$

Table 6.7 All possible realization of the decision intervals in presented in Table 5.6 for Type IV group.

b	c	a	a	c	b	a	b	c
{0,0}	{0,0}	a=0	{0,0}	{0,0}	b=0	{0,0}	{0,0}	1⊇c⊇0
{0,0}	{0,1}	a={1,0}	{0,0}	{0,1}	b={1,0}	{0,0}	{0,1}	1⊇c⊇0
{0,0}	{1,0}	a={0,1}	{0,0}	{1,0}	b={0,1}	{0,0}	{1,0}	1⊇c⊇0
{0,0}	{1,1}	a=1	{0,0}	{1,1}	b=1	{0,0}	{1,1}	1⊇c⊇0
{1,0}	{0,0}	{1,0}⊇a⊇0	{1,0}	{0,0}	{1,0}⊇b⊇0	{1,0}	{0,0}	1⊇c⊇0
{1,0}	{1,0}	a={1,0}	{1,0}	{1,0}	b={1,0}	{1,0}	{1,0}	1⊇c⊇{1,0}
{1,0}	{0,1}	1⊇a⊇{0,1}	{1,0}	{0,1}	b={0,1}	{1,0}	{0,1}	1⊇c⊇0
{1,0}	{1,1}	a=1	{1,0}	{1,1}	b=1	{1,0}	{1,1}	1⊇c⊇{1,0}
{0,1}	{0,0}	{0,1}⊇a⊇0	{0,1}	{0,0}	{0,1}⊇b⊇0	{0,1}	{0,0}	1⊇c⊇0
{0,1}	{1,0}	1⊇a⊇{1,0}	{0,1}	{1,0}	1⊇b⊇{1,0}	{0,1}	{1,0}	1⊇c⊇0
{0,1}	{0,1}	a={0,1}	{0,1}	{0,1}	b={0,1}	{0,1}	{0,1}	1⊇c⊇{0,1}
{0,1}	{1,1}	a=1	{0,1}	{1,1}	b=1	{0,1}	{1,1}	1⊇c⊇{0,1}
{1,1}	{0,0}	1⊇a⊇0	{1,1}	{0,0}	1⊇b⊇0	{1,1}	{0,0}	1⊇c⊇0
{1,1}	{1,0}	1⊇a⊇{1,0}	{1,1}	{1,0}	1⊇b⊇{1,0}	{1,1}	{1,0}	1⊇c⊇{1,0}
{1,1}	{0,1}	1⊇a⊇{0,1}	{1,1}	{0,1}	1⊇b⊇{0,1}	{1,1}	{0,1}	1⊇c⊇{0,1}
{1,1}	{1,1}	a=1	{1,1}	{1,1}	b=1	{1,1}	{1,1}	c=1

Applications of the Reflexive Game Theory: Advanced Topics

The decision interval for subject a, b and c are presented in Table 6.7. Using this table, it is possible to construct the decision and probability spectra for all the subjects.

The decision spectrum for subject a is
$P(1 \supseteq a \supseteq 0) = 1/16$;
$P(1 \supseteq a \supseteq \{0,1\}) = 2/16$;
$P(1 \supseteq a \supseteq \{1,0\}) = 2/16$;
$P(\{0,1\} \supseteq a \supseteq 0) = 1/16$;
$P(\{1,0\} \supseteq a \supseteq 0) = 1/16$;
$P(a = 1) = 4.14$;
$P(a = \{1,0\}) = 2/16$;
$P(a = \{0,1\}) = 2/16$;
$P(a = 0) = 1/16$.

The decision spectrum for subject b is the same as the one for subject a.

The decision spectrum for subject c is
$P(1 \supseteq c \supseteq \{0,1\}) = 3/16$;
$P(1 \supseteq c \supseteq \{1,0\}) = 3/16$;
$P(1 \supseteq c \supseteq 0) = 9/16$;
$P(c=1) = 1/16$.

The probability spectra of alternatives for subject a is
$P_m(a=0) =$
$P(a=0) + P(a=0|1 \supseteq a \supseteq 0) +$
$P(a=0|\{0,1\} \supseteq a \supseteq 0) + P(a=0|\{1,0\} \supseteq a \supseteq 0) =$
$1/16 + 1/16 \cdot 1/4 + 1/16 \cdot 1/2 + 1/16 \cdot 1/2 = 9/64$.

$P_m(a = \{1,0\}) =$
$P(a=\{1,0\}) + P(a=\{1,0\}|1 \supseteq a \supseteq 0) +$
$P(a=\{1,0\}|1 \supseteq a \supseteq \{1,0\}) + P(a=\{1,0\}|\{1,0\} \supseteq a \supseteq 0) =$
$2/16 + 1/16 \cdot 1/4 + 2/16 \cdot 1/2 + 1/16 \cdot 1/2 = 15/64$.

$P_m(a = \{0,1\}) =$

$P(a=\{0,1\}) + P(a=\{0,1\}|1\supseteq a\supseteq 0) +$
$P(a=\{0,1\}|1\supseteq a\supseteq\{0,1\}) + P(a=\{0,1\}|\{0,1\}\supseteq a\supseteq 0) =$
$2/16 + 1/16\cdot 1/4 + 2/16\cdot 1/2 + 1/16\cdot 1/2 = 15/64$.

$P_m(a=1) =$
$P(a=1) + P(a=1|1\supseteq a\supseteq 0) +$
$P(a=1|1\supseteq a\supseteq\{1,0\}) + P(a=1|1\supseteq a\supseteq\{0,1\}) =$
$4.14 + 1/16\cdot 1/4 + 2/16\cdot 1/2 + 2/16\cdot 1/2 = 25/64$.

The same probability spectrum is for subject b.

Probability spectrum for subject c is
$P_m(c=0) = P(c=0|1\supseteq c\supseteq 0) = 9/16\cdot 1/4 = 9/64$.

$P_m(c=\{1,0\}) = P(c=\{1,0\}|1\supseteq c\supseteq 0) + P(c=\{1,0\}|1\supseteq c\supseteq\{1,0\}) =$
$9/16\cdot 1/4 + 3/16\cdot 1/2 = 15/64$.

$P_m(c=\{0,1\}) = P(c=\{0,1\}|1\supseteq c\supseteq 0) + P(c=\{0,1\}|1\supseteq c\supseteq\{1,0\}) =$
$9/16\cdot 1/4 + 3/16\cdot 1/2 = 15/64$.

$P_m(c=1) = P(c=1) + P(c=1|1\supseteq c\supseteq 0) +$
$P(c=1|1\supseteq c\supseteq\{1,0\}) + P(c=1|1\supseteq c\supseteq\{0,1\}) =$
$1/16 + 9/16\cdot 1/4 + 3/16\cdot 1/2 + 3/16\cdot 1/2 = 25/64$.

Surprisingly probability spectrum is the same all three subjects.

6.5 Modeling Social Dynamics with Markov Stochastic Process

In the previous sections, we have computed the probability spectra of alternatives for subjects in all eight groups of the groups.

The essence of the group dynamics is to model how groups can change their structure. Given a certain group structure, it is possible to compute probabilities of how the group can change its structure.

In Figs. 6.2 and 6.3, we introduce transition probabilities of groups of types III and IV.

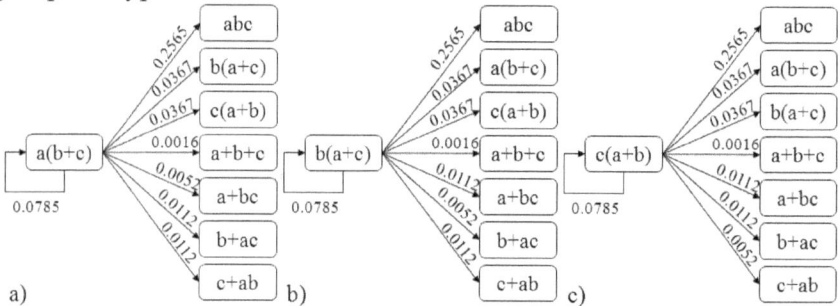

Fig. 6.2 Transition probabilities for groups of Type III.

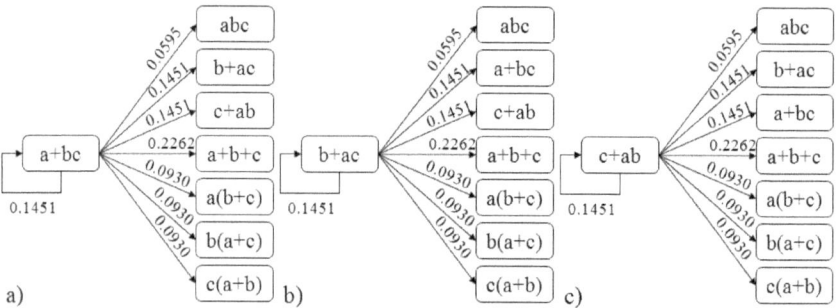

Fig. 6.3 Transition probabilities for groups of Type IV.

Table 6.8 Transition probability matrix Λ

	abc	a+b+c	a+bc	b+ac	c+ab	a(b+c)	b(a+c)	c(a+b)
abc	1.0000	0.0000	0.0000	0.0000	0.0000	0.0000	0.0000	0.0000
a+b+c	1.0000	0.0000	0.0000	0.0000	0.0000	0.0000	0.0000	0.0000
a(b+c)	0.2565	0.0016	0.0052	0.0112	0.0112	0.0785	0.0367	0.0367
b(a+c)	0.2565	0.0016	0.0112	0.0052	0.0112	0.0367	0.0785	0.0367
c(a+b)	0.2565	0.0016	0.0112	0.0112	0.0052	0.0367	0.0367	0.0785
a+bc	0.0595	0.2262	0.1451	0.1451	0.1451	0.0930	0.0930	0.0930
b+ac	0.0595	0.2262	0.1451	0.1451	0.1451	0.0930	0.0930	0.0930
c+ab	0.0595	0.2262	0.1451	0.1451	0.1451	0.0930	0.0930	0.0930

It is also important to note that groups of type can have one subject in a frustration state. A probability of the frustration state is 0.5625. Once in frustration state, subject cannot make any decision and group collapses. Since group structure can be change if and only all the subject can make their decision, groups of type III can change their structure only with probability 0.4375. Therefore, for groups of Type III all transition probabilities sum up to 0.4375 not to 1.0000.

Since, there is no any possibility of frustration for subjects in groups of type IV, transition probabilities of groups of type IV sum up to 1.000.

Given the transition probabilities, how can we model group dynamics? The RTG inference guarantees that during the decision making only the present state of the group and the influences of the subjects on each other effect the actual decision.

Table 6.9 Transition probability matrix Λ

	abc	a+b+c	a+bc	b+ac	c+ab	a(b+c)	b(a+c)	c(a+b)
Abc	1.0000	0.0000	0.0000	0.0000	0.0000	0.0000	0.0000	0.0000
a+b+c	1.0000	0.0000	0.0000	0.0000	0.0000	0.0000	0.0000	0.0000
a(b+c)	0.5863	0.0037	0.0119	0.0256	0.0256	0.1794	0.0838	0.0838
b(a+c)	0.5863	0.0037	0.0256	0.0119	0.0256	0.0838	0.1794	0.0838
c(a+b)	0.5863	0.0037	0.0256	0.0256	0.0119	0.0838	0.0838	0.1794
a+bc	0.0595	0.2262	0.1451	0.1451	0.1451	0.0930	0.0930	0.0930
b+ac	0.0595	0.2262	0.1451	0.1451	0.1451	0.0930	0.0930	0.0930
c+ab	0.0595	0.2262	0.1451	0.1451	0.1451	0.0930	0.0930	0.0930

This illustrates the Markov property, when any at any state the future state of the system depends only on the present state. Therefore, a group dynamics is Markov Stochastic Process and can be modeled using transition probability matrix Λ. Such matrix is presented in Table 6.8. The problem is that for groups of Type III, the overall sum is 0.4375.

Table 6.10 Transition probability matrix Λ^2

	abc	a+b+c	a+bc	b+ac	c+ab	a(b+c)	b(a+c)	c(a+b)
abc	1.0000	0.0000	0.0000	0.0000	0.0000	0.0000	0.0000	0.0000
a+b+c	1.0000	0.0000	0.0000	0.0000	0.0000	0.0000	0.0000	0.0000
a+bc	0.6476	0.0787	0.0518	0.0516	0.0516	0.0387	0.0400	0.0400
b+ac	0.6476	0.0787	0.0516	0.0518	0.0516	0.0400	0.0387	0.0400
c+ab	0.6476	0.0787	0.0516	0.0516	0.0518	0.0400	0.0400	0.0387
a(b+c)	0.5575	0.0647	0.0496	0.0496	0.0496	0.0763	0.0763	0.0763
b(a+c)	0.5575	0.0647	0.0496	0.0496	0.0496	0.0763	0.0763	0.0763
c(a+b)	0.5575	0.0647	0.0496	0.0496	0.0496	0.0763	0.0763	0.0763

Table 6.11 Transition probability matrix Λ^3

	abc	a+b+c	a+bc	b+ac	c+ab	a(b+c)	b(a+c)	c(a+b)
abc	1.0000	0.0000	0.0000	0.0000	0.0000	0.0000	0.0000	0.0000
a+b+c	1.0000	0.0000	0.0000	0.0000	0.0000	0.0000	0.0000	0.0000
a+bc	0.8242	0.0274	0.0205	0.0205	0.0205	0.0290	0.0290	0.0290
b+ac	0.8242	0.0274	0.0205	0.0205	0.0205	0.0290	0.0290	0.0290
c+ab	0.8242	0.0274	0.0205	0.0205	0.0205	0.0290	0.0290	0.0290
a(b+c)	0.7231	0.0523	0.0363	0.0363	0.0363	0.0385	0.0385	0.0385
b(a+c)	0.7231	0.0523	0.0363	0.0363	0.0363	0.0385	0.0385	0.0385
c(a+b)	0.7231	0.0523	0.0363	0.0363	0.0363	0.0385	0.0385	0.0385

Table 6.12 Transition probability matrix Λ^{10}

	abc	a+b+c	a+bc	b+ac	c+ab	a(b+c)	b(a+c)	c(a+b)
abc	1.0000	0.0000	0.0000	0.0000	0.0000	0.0000	0.0000	0.0000
a+b+c	1.0000	0.0000	0.0000	0.0000	0.0000	0.0000	0.0000	0.0000
a+bc	0.9962	0.0007	0.0005	0.0005	0.0005	0.0006	0.0006	0.0006
b+ac	0.9962	0.0007	0.0005	0.0005	0.0005	0.0006	0.0006	0.0006
c+ab	0.9962	0.0007	0.0005	0.0005	0.0005	0.0006	0.0006	0.0006
a(b+c)	0.9944	0.0010	0.0007	0.0007	0.0007	0.0008	0.0008	0.0008
b(a+c)	0.9944	0.0010	0.0007	0.0007	0.0007	0.0008	0.0008	0.0008
c(a+b)	0.9944	0.0010	0.0007	0.0007	0.0007	0.0008	0.0008	0.0008

Therefore, we normalized transition probabilities by dividing it by 0.4375. A new transition probability matrix is presented in Table 6.9. This new transition probability matrix Λ is a matrix of transition probabilities, if no frustration occurs during decision making.

Table 6.13 Transition probability matrix Λ^{20}

	abc	a+b+c	a+bc	b+ac	c+ab	a(b+c)	b(a+c)	c(a+b)
Abc	1.0000	0.0000	0.0000	0.0000	0.0000	0.0000	0.0000	0.0000
a+b+c	1.0000	0.0000	0.0000	0.0000	0.0000	0.0000	0.0000	0.0000
a+bc	1.0000	0.0000	0.0000	0.0000	0.0000	0.0000	0.0000	0.0000
b+ac	1.0000	0.0000	0.0000	0.0000	0.0000	0.0000	0.0000	0.0000
c+ab	1.0000	0.0000	0.0000	0.0000	0.0000	0.0000	0.0000	0.0000
a(b+c)	1.0000	0.0000	0.0000	0.0000	0.0000	0.0000	0.0000	0.0000
b(a+c)	1.0000	0.0000	0.0000	0.0000	0.0000	0.0000	0.0000	0.0000
c(a+b)	1.0000	0.0000	0.0000	0.0000	0.0000	0.0000	0.0000	0.0000

In matrix Λ elements of each row sum up to 1.000. Therefore, using matrix Λ we can model group dynamics or social dynamics by applying powers of matrix Λ and compute transition probabilities on each step of Markov process. The transition probabilities after n iterations are modeled as nth power of transition probability matrix Λ: $\Lambda \to \Lambda^{n+1}$. Here we present several iterations of the group dynamics. Tables 6.10-6.13 illustrate results transition probability matrices after the 1st, 2nd, 9th and 19th iterations.

Table 6.14 Transition probability matrix Λ_{Type}

	Type I	Type II	Type III	Type IV
Type I	1.0000	0.0000	0.0000	0.0000
Type II	1.0000	0.0000	0.0000	0.0000
Type III	0.5863	0.0037	0.0631	0.3470
Type IV	0.0595	0.2262	0.4353	0.2790

It is easy to see that during the social dynamics the transition probabilities for groups of type III and IV to change into type I group constantly increase. After 19th iteration (Table 6.13).

Table 6.15 Transition probability matrix Λ^2_{Type}

	Type I	Type II	Type III	Type IV
Type I	1.0000	0.0000	0.0000	0.0000
Type II	1.0000	0.0000	0.0000	0.0000
Type III	0.6476	0.0787	0.1550	0.1187
Type IV	0.5575	0.0647	0.1489	0.2289

Table 6.16 Transition probability matrix Λ^4_{Type}

	Type I	Type II	Type III	Type IV
Type I	1.0000	0.0000	0.0000	0.0000
Type II	1.0000	0.0000	0.0000	0.0000
Type III	0.8928	0.0199	0.0417	0.0456
Type IV	0.8462	0.0265	0.0572	0.0701

Table 6.17 Transition probability matrix Λ^8_{Type}

	Type I	Type II	Type III	Type IV
Type I	1.0000	0.0000	0.0000	0.0000
Type II	1.0000	0.0000	0.0000	0.0000
Type III	0.9885	0.0020	0.0043	0.0051
Type IV	0.9831	0.0030	0.0064	0.0075

Table 6.18 Transition probability matrix Λ^{16}_{Type}

	Type I	Type II	Type III	Type IV
Type I	1.0000	0.0000	0.0000	0.0000
Type II	1.0000	0.0000	0.0000	0.0000
Type III	0.9998	0.0000	0.0001	0.0001
Type IV	0.9998	0.0000	0.0001	0.0001

Figs 6.4 and 6.5 illustrate how transition probabilities for groups of types III and IV, respectively, evolve along the iterations

of social interaction and this illustrate social dynamics of group structures.

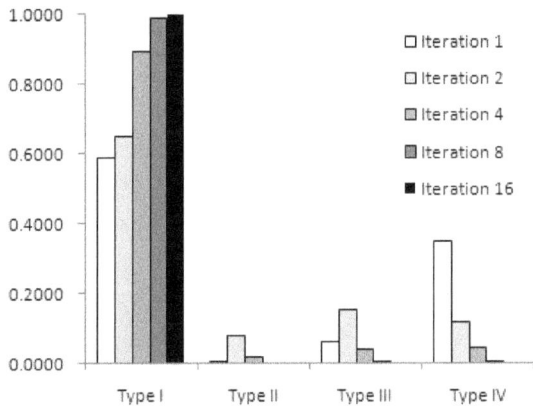

Fig. 6.4 Evolution of transition probabilities of Type III groups.

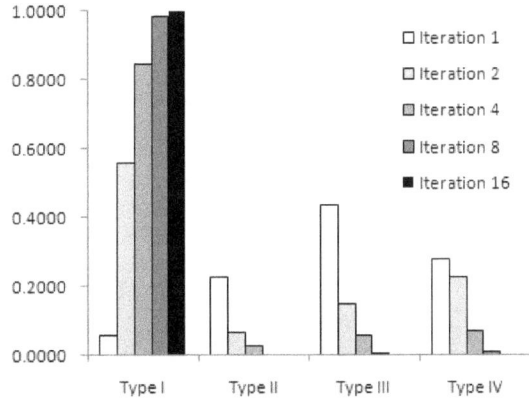

Fig. 6.5 Evolution of transition probabilities of Type IV groups.

Finally, we address the issue of marginal transition probabilities from one type of groups to another one. A new matrix Λ_{Type} is obtained from original transition probability matrix Λ by summing up the transition probabilities for all groups in corresponding types.

The transition probability matrices for transitions between types are presented in on various iterations are presented in Tables 6.14-6.18.

Finally, we call a Markov Process based on probabilities derived from the RGT inference to be a *Reflexive Markov Process*.

Notes

References

Adams-Webber, F. (1978) A further test of golden section hypothesis. *British Journal of Psychology*, 69, pp. 439-442.

Asimov, I. (1942) Runaround. *Astounding Science Fiction*, March, pp. 94-103.

Benjafield, F., and Adams-Webber, F. (1976) The golden section hypothesis. *British Journal of Psychology*, 67, pp. 11-15.

Benjafield, F. and Green, T.R.G. (1978) Golden section relation in interpersonal judgment. *British Journal of Psychology*, 69, pp. 25-35.

Batchelder, W.H., and Lefebvre, V.A. (1982) A mathematical analysis of a natural class of partitions of a graph. *Journal Mathematical Psychology*, 26, pp. 124-148.

Hopkins, M.S. (2011) Big Data, Analytics and the Path from Insights to Value: From Editor, *MIT Sloman Management Review*, 52, 2, pp. 21-32.

Ekman, P. (2007) *Emotions Revealed: Recognizing Faces and Feelings to Improve Communication and Emotional Life*. Holt Paperbacks.

Ekman, P., and Davidson, R. J. (1993) Voluntary changes regional brain activity. *Psychological Science*, 4, 5, pp. 34.2045.

Levenson, R. W., Ekman, P., and Friesen, W. V. (1990) Voluntary facial action generates emotion specific autonomic nervous system activity. *Psychophysiology*, 27 (4), pp. 363-384.

Izhikevich, E. M. (2001) Resonate-and-fire neurons. *Neural Networks*, 14, pp. 883-894.

Johnson, D. E. A. (2010) Human socio-cultural behavior studies: three simple questions. *Defense Concepts by Center for Advanced Defense. Studies*, 5, 4, pp. 47-57.

Kelly, G.A. (1955) *The psychology of personal constructs*. New York: Norton.

Lefebvre, V.A. (1965) The basic ideas of reflexive game's logic. *Problems of research of systems and structures*. pp. 73-79. [in Russian]

Lefebvre V.A. (1977) *The Structure of Awareness: Toward a Symbolic Language of Human Awareness*. Beverly Hills, Calif.: Sage.

Lefebvre, V.A. (1982) *Algebra of Conscience*. D. Reidel, Holland.

Lefebvre, V.A. (1985) The Golden section and an algebraic model of ethical cognition. *Journal of Mathematical Psychology*, 29, pp. 289-310.

Lefebvre, V.A. (2001) *Algebra of Conscience. 2nd Edition*. Holland: Kluwer.

Lefebvre, V.A. (2006) *Research on Bipolarity and Reflexivity*. Edwin Mellen, Ney York.

Lefebvre, V.A. (2009) *Lecture notes on the Reflexive Game Theory*. Cognito-Center, Moscow. [in Russian]

Lefebvre, V.A. (2010) *Lecture on Reflexive Game Theory*. Leaf & Oaks, Los Angeles.

Mehrabian, A. (1996) Pleasure-arousal-dominance: a general framework for describing and measuring individual differences in temperament. Current Psychology: Developmental, Learning, Personality, Social, 14, 4.231-292.

Osgood, C.E. (1979) From Yang and Yin to AND and BUT in cross cultural perspective. *International Journal of Psychology*, 14, pp. 1-35.

Osgood, C.E., and Richards, M.M. (1973) From Yang and Yin to AND and BUT. *Language*, 49, 1, pp. 380-4.19.

Osgood, C.E., Suci, G., and Tannenbaum, P. (1957) *The measurement of meaning*. Urbana, IL: University of Illinois Press.

Russell, J.A., and Mehrabian, A. (1977) Evidence for a three-factor theory of emotions. Journal of Research in Personality 11, pp. 273-294.

Shalit, B. (1980) The golden section relation in the evaluation of environment factors. British Journal of Psychology, 71, pp. 39-42.

Tarasenko, S. (2010) The Inverse Task of the Reflexive Game Theory: Theoretical Matters, Practical Applications and Relationship with Other Issues. arXiv:1011.3397 [cs.MA]

Tarasenko, S. (2011) Modeling mixed groups of humans and robots with Reflexive Game Theory. In Lamers, M.H., and Verbeek, F.J. (eds.): HRPR 2010, LINCST 59, pp. 108-117.

Tarasenko, S. (2013) Modeling multi-stage decision processes with Reflexive Game Theory. Studia Humana, 2(3), pp. 46-52.

Tarasenko, S., and Inui, T. (2009) Blind Choice, *Perceptual and Motor Skills*, 109, 3, 791-803 (2009).

Tarasenko, S., Inui, T., and Abdikeev, N.M. (2006) Non-random Human Performance under conditions of Lacking Information. In Proceedings for 28th Annual Conference of the Cognitive Science Society, Vancouver, Canada, July 26-29, pp. 2228-2233. Erlbaum: NJ.

Yani-de-Soriano, M.M., and Foxall, G.R. (2006) The emotional power of place: The fall and rise of dominance in retail research. Journal of Retailing and Consumer Services, 13, pp. 403-4.14.

www.ingramcontent.com/pod-product-compliance
Lightning Source LLC
Chambersburg PA
CBHW060903170526
45158CB00001B/476